SPINAL CORD PLASTICITY

Alterations in Reflex Function

SPINAL CORD PLASTICITY

Alterations in Reflex Function

edited by

Michael M. Patterson
Nova Southeastern University

James W. Grau
Texas A & M University

KLUWER ACADEMIC PUBLISHERS
Boston / Dordrecht / London

Distributors for North, Central and South America:
Kluwer Academic Publishers
101 Philip Drive
Assinippi Park
Norwell, Massachusetts 02061 USA
Telephone (781) 871-6600
Fax (781) 681-9045
E-Mail <kluwer@wkap.com>

Distributors for all other countries:
Kluwer Academic Publishers Group
Distribution Centre
Post Office Box 322
3300 AH Dordrecht, THE NETHERLANDS
Telephone 31 78 6392 392
Fax 31 78 6546 474
E-Mail <services@wkap.nl>

Electronic Services <http://www.wkap.nl>

Library of Congress Cataloging-in-Publication Data
Spinal cord plasticity: alterations in reflex function/ edited by Michael M. Patterson, James W. Grau.
p.; cm.
Includes bibliographical references and index.
ISBN 0-7923-7445-2 (alk. paper)
1. Spinal cord. 2. Neuroplasticity. 3. Reflexes. I. Patterson, Michael M. II. Grau, James W.
[DNLM: 1. Spinal Cord—physiology. 2. Neuronal Plasticity—physiology. 3. Reflex—physiology. 4. Spinal Cord Injuries—rehabilitation. WL 400 S7583 2001]
QP371 .S645 2001
612.8'3—dc21

2001038130

Printed on acid-free paper. Printed in the United States of America

TABLE OF CONTENTS

Foreword

The area of spinal cord plasticity has become a very actively researched field. The spinal cord has long been known to organize reflex patterns and serve as the major transmission pathway for sensory and motor nerve impulses. However, the role of the spinal cord in information processing and in experience driven alterations is generally not recognized. With recent advances in neural recording techniques, behavioral technologies and neural tracing and imaging methods has come the ability to better assess the role of the spinal cord in behavioral control and alteration. The discoveries in recent years have been revolutionary. Alterations due to nociceptive inputs, simple learning paradigms and repetitive inputs have now been documented and their mechanisms are being elucidated.

These findings have important clinical implications. The development of pathological pain after a spinal cord injury likely depends on the sensitization of neurons within the spinal cord. The capacity of the spinal cord to change as a function of experience, and adapt to new environmental relations, also affects the recovery locomotive function after a spinal cord injury. Mechanisms within the spinal cord can support stepping and the capacity for this behavior depends on behavioral training. By taking advantage of the plasticity inherent within the spinal cord, rehabilitative procedures may foster the recovery of function.

The chapters in this book explore some of these recent advances in spinal cord plasticity. The authors characterize spinal cord plasticity from both a behavioral and molecular perspective and describe how behavioral and pharmacological treatments can foster the recovery of function after a spinal cord injury.

The book is the outgrowth of a symposium organized by the editors and Reggie Edgerton, Jonathan Wolpaw and William Willis. The symposium was held at the UCLA Tennis Center in conjunction with the Society for Neuroscience Meetings on November 7, 1998. The organizers wish to thank the sponsors of the symposium for their support and for the use of the UCLA facilities. The program was sponsored by the Parke-Davis division of Warner-Lambert, the Kent Waldrep Foundation, the Wadsworth Center (Albany), and the Department of Physiological Science, the Neural Repair Program and the Brain Research Institute **at** UCLA. A special thanks goes to Linda Maninger in helping to organize the facilities for the symposium at UCLA.

We also want to thank Brenda Riepenhof for her work on the preparation of the final manuscripts of the book chapters. Our thanks also to Stephanie Washburn at Texas A and M and Stephanie Elkins at UCLA for their most gracious help during the editing process. Our appreciation goes to Michael Williams and Mary Panarelli for their continuing and unwavering help in the publishing process.

Michael M. Patterson

James W. Grau

1 SPINAL PLASTICITY

Richard F. Thompson
Neuroscience Program
University of Southern California
University Park
Los Angeles, CA 90089-2520

"Traditional views often held that little learning occurs within the spinal cord, however, recent findings suggest that neurons within the spinal cord are highly plastic" (Organizers' statement for the symposium).

1.1 SPINAL HABITUATION

One aspect of spinal plasticity with a long history is habituation of spinal reflexes. Sherrington (1906) studied this phenomenon at length in the spinal dog. He used the term fatigue but was careful to define it only as the response decrement to repeated stimulation. Indeed, in ingenious experiments he ruled out sensory adaptation and muscle fatigue as contributing factors and presented strong evidence from reflex interaction studies that the final common path, the spinal motor neurons were not themselves fatigued.

In a series of studies by the author and William Alden Spencer, we characterized the processes of habituation and sensitization in detail at both behavioral and neuronal levels (Thompson and Spencer, 1966). We used habituation and sensitization of the hindlimb flexion reflex in the spinal animal (cat) as a model system to explore the neuronal mechanisms underlying this form of behavioral plasticity, often now termed non-associative learning. Kandel and associates completed similar studies on mechanisms of habituation and sensitization in a monosynaptic pathway in the marine mollusk *Aplysia* (Carew, Pinsker and Kandel, 1972; Kandel, 1976), as we also did using a monosynaptic pathway in the isolated frog spinal cord (Farel and Thompson, 1976; Farel, Glanzman and Thompson, 1973). A primary goal in all these studies was to understand the mechanisms of these forms of behavioral plasticity at the neuronal/synaptic/ molecular levels.

Rather than review these now "historical" studies in any detail I summarize here the results of our analytic studies on localization and identification of the process of plasticity in the mammalian spinal cord. A schematic of the spinal cord is shown in Figure 1.1, together with the experimental manipulations used. The repeated stimulus was to skin (Ss) or cutaneous nerve (Sin) and the behavioral response was contraction of a flexor muscle (Rm), recording of the ventral root response (Rn) or intracellular recording from the alpha motor neurons (MRa).

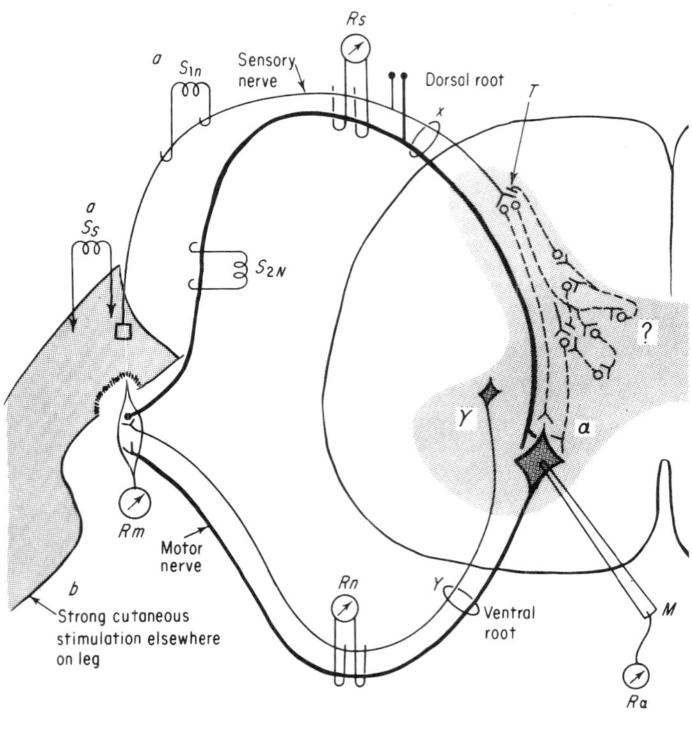

Figure 1.1. Much simplified schematic of the mammalian spinal cord indicating some of the experimental procedures used to study plasticity (habituation, sensitization and classical conditioning). Electrical stimuli can be delivered to skin (Ss, b), to afferent nerves (Sin, cutaneous nerve; S_2N, muscle afferent nerve; x, dorsal root) or to cutaneous nerve terminals for antidromic activation (T). Responses can be recorded from a flexor muscle (Rm), a motor nerve (Rn), the ventral root (Y), intracellularly from the motor neurons (MRa) or antidromically from the mixed or cutaneous nerve (Rs). See text for more details. (From Thompson, 1967).

Since both skin (Ss) and sensory nerve stimulation (Sin) yielded identical habituation, sensory adaptation was ruled out. Since both muscle contraction (Rm) and ventral root response (Rn) showed identical habituation, muscle fatigue was ruled out. The possibility that motor neuron excitability decreased with repeated stimulation was evaluated by interpolated stimulation of a muscle nerve (S_2N) activating the monosynaptic Ia afferent to motor neuron response. This monosynaptic response, recorded either from the ventral root (Rm) or intracellularly from the motor neuron (MRa), did not change at all over the course of habituation. Similarly, spike generation threshold to intracellular stimulation via the intracellular electrode (MRa) did not change during habituation. The possibility that alterations in cutaneous afferent terminal excitability, e.g., presynaptic inhibition (at T in Figure 1.1) was tested by interpolated stimulation of the afferent terminals at T with antidromic recording of the dorsal root response at Rs. This antidromic dorsal root response did not change at all over the course of habituation to repeated skin or cutaneous nerve stimulation, thus ruling out changes in terminal excitability.

Sensitization in this system was induced by strong cutaneous stimulation elsewhere on the leg (see Figure 1.1). This strong extra stimulation caused an increase in the habituated response, and in the control response to cutaneous stimulation before habitation training. Unlike habituation, sensitization resulted in a significant increase in excitability in motor neurons. Thus, following the sensitizing stimulus there was a significant increase the monosynaptic muscle nerve-motor neuron response.

These experiments localized the process of habituation to spinal interneurons and suggested that interneurons played a key role in both habituation and sensitization.

In studies with Philip Groves, recording from spinal interneurons during habituation and sensitization, we identified several classes of interneurons, including one class that showed only decrements to repeated stimulation, independent of reflex sensitization and another class that showed dramatic increases in response during sensitization. Thus led us to the "dual process" theory of habituation (Groves and Thompson, 1970).

The basic idea is that repeated stimulation resulted in synaptic processes of decrement and increment and the two processes interact to yield the behavioral outcome. This is schematized in oversimplified terms in Figure 1.1, where the decremental process would occur in the interneurons between cutaneous afferents and alpha motor neurons (shown as a single interneuron in Figure 1.1), whereas the process of sensitization resulted in a build-up of excitation in an interneuron pool (? in Figure 1.1), leading to increased excitability of the motor neuron at *a*. We were able to account for a wide range of behavioral (and neuronal) phenomena of habituation and sensitization with this theory. In monosynaptic systems, where appropriate analysis is possible, e.g., Aplysia and isolated frog spinal cord, habituation is a presynaptic process apparently involving a decrease in probability of transmitter release as a result of repeated stimulation.

1.2 SPINAL CONDITIONING

Unlike habituation, the early history of spinal conditioning -- the possibility that classical or instrumental training procedures could induce associative learning - like phenomena in the spinal cord -- was somewhat controversial. Pavlov's dictum to the effect that associative learning required the cerebral cortex did not help matters.

Shurrager, working in Culler's laboratory (where so many pioneering studies of brain substrates of learning and memory were carried out), published the first modern studies of classical conditioning of spinal reflexes, (Shurrager and Culler, 1940; 1941). In brief, they used acute spinal dogs, measured the twitch response of a partially dissected flexor muscle, gave paw shock as a US and weak stimulation of the tail as CS. They obtained robust acquisition in about half their animals and demonstrated CS alone extinction and successively more rapid reacquisition in repeated training and extinction sessions. Unfortunately, adequate controls for sensitization and pseudoconditioning were not run in these studies.

A few years later Kellogg and associates reported negative results in attempts at spinal conditioning (e.g., Kellogg, 1947; Kellogg et. al., 1946; Deese and Kellogg, 1949). They used chronic spinal dogs and the flexor response of the whole leg. The US was shock to the paw of that leg and the CS was a shock to the opposite hind paw. Kellogg's choice of CS locus was unfortunate. Paw shock elicits a crossed extension reflex that would work against the development of a conditioned flexion response. Pinto and Bromiley (1950) completed an extensive spinal conditioning study with long-term acute spinal animals and found only inconclusive evidence because of passive hindquarter movements caused by anterior limb movements.

The bible in the field at that time (Morgan and Stellar, *Physiological Psychology*, 2nd ed., 1950) drew the following conclusion:

> "So there is a good deal wrong with the experiments as they stand. They do not let us conclude confidently that spinal conditioning can take place. We need more experimenting before we can be sure. In fact, it may turn out that what seemed to be conditioning is more correctly called reflex sensitization, as Kellogg has suggested above. Even if that is the case the experiments will be of some value, for knowing as little as we do about conditioning, it is possible that conditioning may be closely related to reflex sensitization." (p. 446)

This conclusion apparently discouraged others from pursuing the phenomena of spinal conditioning for many years to come.

In 1967 the author and a graduate student took up the gauntlet (Fitzgerald and Thompson, 1967). We used acute spinal cats, detached the distal tendon of left hind limb flexor muscle (tibialis anterior) led it out through a small skin incision and attached it to a force-displacement transducer to measure the flexor responses. The CS was a weak shock to thigh skin and the US was a brief strong shock to the left hind paw, CS and US coterminating. All animals were first given four UCS alone trials following by a series of CS alone trials to establish a control response level. Half the animals were then given paired CS-US trials and half were given explicitly unpaired CS and US trials. Both groups were then given CS alone extinction training. The paired groups showed a rapid initial increase in response to the CS. Since these were pilot studies, we varied the CS-UCS onset interval, number of training trials and CS and US shock parameters. Consequently, the overall results should be viewed with caution. Our impressions were that most of the increase in the paired group occurred in the first few trials and a 0.5 sec CS-US onset interval seemed more effective than a 0.2 sec interval.

Subsequently, we completed a detailed and extensive study of spinal conditioning, using a number of control procedures and conditions, which yielded clear positive results (Patterson, Cegavske and Thompson, 1973). We ran a number of pilot animals (cats) to establish the most effective conditions: animals were anesthetized, spinalized (T-12), given local anesthetics, then paralyzed with Flaxedil and given artificial respiration. The superficial and deep peroneal motor nerves were dissected out and placed on stimulating and recording electrodes. The CS was a weak shock (eight 1 ms pulses at 10 Hz) to the superficial peroneal response of intensity yielding a motor response to the first pulse of 0.02 to 0.04 mV. The US was a series of pulses (thirteen 2 ms, 50v 25 Hz) to skin to the left ankle, yielding a UR of at least 0.08 mV. All groups first received a series of 15 CS alone trials. The conditioning group received 75 acquisitive trials, 250 msec forward ISI, and 50 CS alone extinction trials. Control groups received explicitly unpaired CS and US trials (75 each). In one series a CS alone trials group was also included. Two separate experiments were completed; both showed clear evidence of associative learning. Results of the second experiment are shown in Figure 1.2.

These experiments clearly ruled out sensitization as a process responsible for the increase in CS response in the paired group. The fact that the animals were paralyzed ruled out movement artifacts. Acquisition was rapid, as was extinction, just as in the original Shurrager and Culler studies and in our earlier study. These results were replicated in careful studies by Durkovic (1975). In a recent and most interesting study, Durkovic and Prokowich (1998) infused intrathecally artificial CSF (controls) or artificial CSF with the NMDA blocker APV during the conditioning period in acute spinal cats, using procedures described above. Both groups showed normal acquisition of the spinal CR. However, the APV group exhibited no retention of the increased response in the 2.5 h retention period, in contrast to the CSF alone group. The results suggest that NMDA receptor activation plays a critical role in the establishment of long-term associative plasticity in the spinal cord.

A key issue for many is the extent to which this form of spinal Pavlovian conditioning resembles Pavlovian conditioning of discrete responses in the intact animal. Patterson and associates completed an heroic series of parametric studies to address this issue, using the same general procedures as Patterson et al. (1973). In brief, spinal conditioning exhibits differential conditioning (Beggs, Steinmetz and Patterson, 1985) forward but not backward conditioning (Patterson 1975), retention of the CR over a period of hours (Beggs et. al., 1983), increasingly effective conditioning with increasing US strength (Polenchar et. al., 1984) and best learning with a 250 onset forward ISI. (See Patterson, 1976, for a detailed review of all studies to that time on spinal conditioning.) All these properties resemble the properties of classical conditioning of discrete responses in intact mammals.

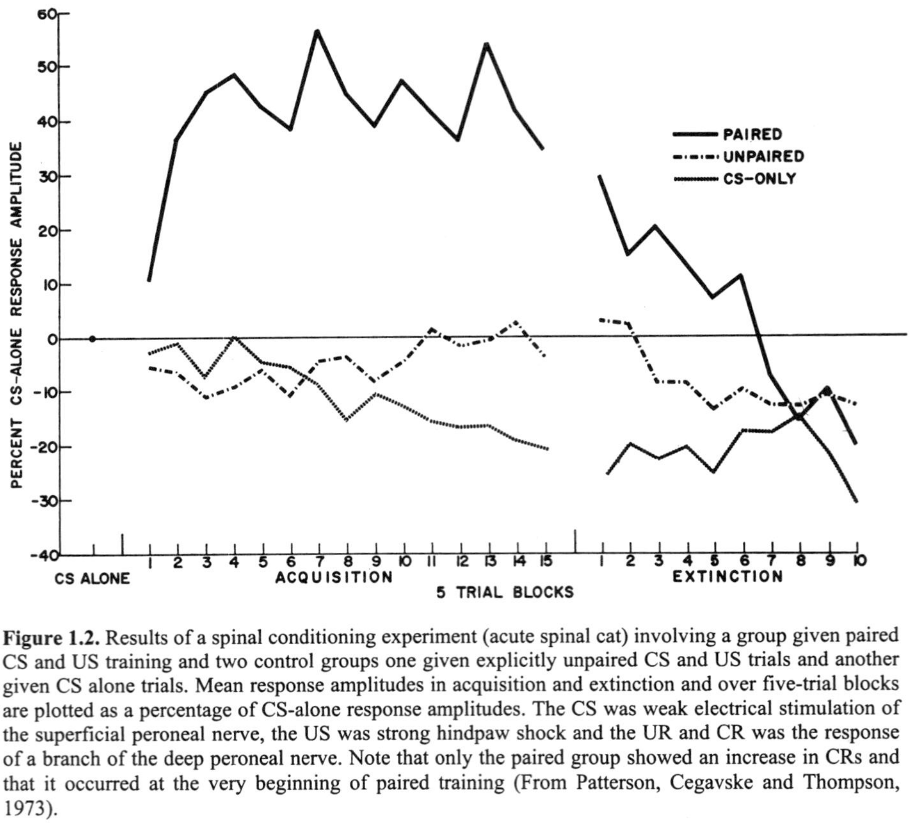

Figure 1.2. Results of a spinal conditioning experiment (acute spinal cat) involving a group given paired CS and US training and two control groups one given explicitly unpaired CS and US trials and another given CS alone trials. Mean response amplitudes in acquisition and extinction and over five-trial blocks are plotted as a percentage of CS-alone response amplitudes. The CS was weak electrical stimulation of the superficial peroneal nerve, the US was strong hindpaw shock and the UR and CR was the response of a branch of the deep peroneal nerve. Note that only the paired group showed an increase in CRs and that it occurred at the very beginning of paired training (From Patterson, Cegavske and Thompson, 1973).

In an interesting recent series of studies Grau and associates paired shock to a hind leg of a CS with *intense* tail shock as a US in the spinal rat and then examined effect of CS presentations on antinociception on the tail-flick test (Joynes & Grau, 1996). This paradigm is complex in that the CR is a variation on the UR. Intense tail shock would seem to induce massive sensitization. Grau interpreted the results, incidentally, as protection from habituation. In any event, the parametric properties of this procedure apparently differ from intact animal conditioning and from the spinal conditioning studies reviewed above. It may therefore not be an appropriate model of Pavlovian conditioning, at least insofar as analytic studies of mechanisms are concerned.

The spinal conditioning results reviewed above resemble classical conditioning of discrete responses in intact animals in many properties, nonetheless appear to differ from intact animal learning in several ways. First, acquisition is very rapid, most increases in response to the CS occurring in the first few trials. Second, and perhaps more important, the onset latency of the CR does not appear to move forward in time over the course of learning (see Figure 1.3 and Patterson & Thompson, unpublished date). Finally, and seemingly most important, spinal conditioning involves an alpha response. In most studies, the CS elicits the UR-CR before training. As a result of training there is an associatively produced increase in the amplitude of the response to the CS. The fact that spinal conditioning may be "alpha" conditioning perhaps accounts for the lack of forward shift of the CR onset with training.

The idea that alpha conditioning differs from normal conditions may be somewhat arbitrary. In one experiment in Patterson et. al., (1973), two branches of the deep peroneal nerve were recorded from during conditioning (Figure 1.3). One branch showed responses to the CS (superficial peroneal nerve stimulus) prior to training but the other branch did not. However, by trial 10 the non-responsive branch did show a response to the CS. Is it the case that one branch of the nerve showed alpha conditioning, whereas the other branch did not exhibit alpha conditioning but rather showed only "normal" conditioning?

In unpublished pilot studies, Patterson and I and associates completed some initial analytic studies in an attempt to localize the site(s) of synaptic plasticity that underlie spinal conditioning, much as Spencer and I had done earlier for habituation and sensitization (see Figure 1.1). The preparation itself, involving cutaneous nerve stimulation (Sin) as the CS, strong cutaneous stimulation of the paw a US and recording motor nerve responses (Rn), ruled out changes in sensory receptors or properties of the muscles. Using the monosynaptic test pathway (S_2N to Rn) we ruled out changes in motor neurons. Similarly changes in the excitability of the cutaneous afferent fibers were ruled out by stimulation of the terminals at T with antidromic recording at Rs. Consequently the mechanisms of synaptic plasticity must reside within the interneuron circuits in the spinal gray.

At that time there were no techniques that would permit definitive localization of the processes of plasticity within these interneuron circuits. Even this ancient and "simple" neuronal system proved intractable to definitive analysis of the mechanisms underlying this form of behavioral plasticity/learning.

What happens in the spinal cord when limb flexion is conditioned in the intact animal? In a long series of studies in recent years, we have utilized classical conditioning of discrete responses in intact mammals (mostly rabbits) to localize the sites of memory storage in the brain. Most of our studies made use of the conditioned eyeblink response but we have also used limb flexion. The bottom line is that the cerebellum and its associated brains stem circuitry are the essential (necessary and sufficient) neuronal system for the learning and memory of these responses. Indeed, evidence is very strong that the memory traces are formed and stored in the cerebellum (see Thompson and Krupa, 1994; Kim and Thompson, 1997 for reviews).

In the otherwise intact animal, lesions in the cerebellar nuclei or rubrospinal tract produce complete and specific abolition of the conditioned limb flexion response (see Krupa & Thompson, 1994). Further, normal animals that undergo leg flexion training prior to spinal transection show no retention or savings of conditioned responses in spinal reflexes following transection (J. Steinmetz, personal communication). The isolated spinal cord is thus capable of mediating a kind of associative neuronal plasticity but does not subserve classical conditioning of the limb flexion response in the intact animal. Spinal conditioning is a useful model to study basic associative plasticity in a simplified neuronal network, but it does not tell us where or how such memories are formed in the intact animal.

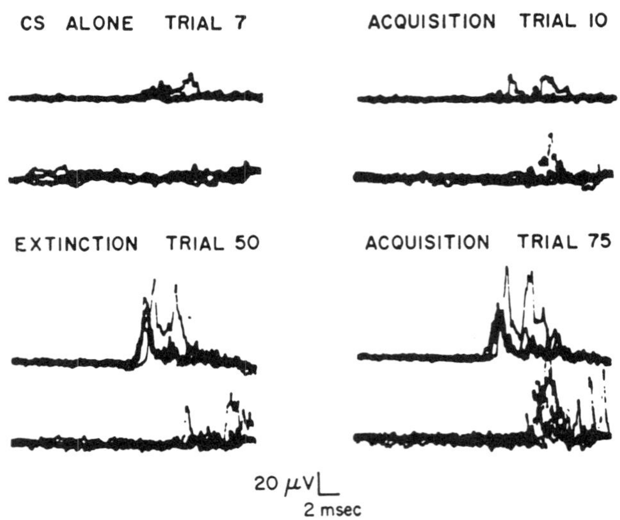

Figure 1.3. Examples of responses of two branches of the deep peroneal nerve on selected trials during the paired conditioning series and CS alone extinction series in one animal (procedures, CS and US as in Figure 6.2). The lower trace of each group is from the branch with the higher CS threshold; the CS could be considered a neutral stimulus for this branch before pairing began. Responses to the first three CS pulses of the CS train (see text) are superimposed in each trace; the stimulus pulse occurs at the beginning of the trace. Upward deflection is negative. (From Patterson, Cegavske and Thompson, 1973).

In the broader context of spinal damage in humans, these data on behavioral and synaptic plasticity in the spinal cord are very promising. The spinal cord seems capable of considerable learning. The ultimate hope is of course to develop methods to regrow or reconnect (with wetware or hardware) severed neuronal tissues in the spinal cord, an area of intense and promising research today.

1.3 REFERENCES

Beggs, A.L., Steinmetz, J.E. and Patterson, M.M. Classical conditioning of a flexor nerve response in spinal cats: Effects of tibial nerve CS and a differential conditioning paradigm. *Behavioral Neuroscience*, 1985; 99:496-508.

Beggs, A.L., Steinmetz, J.E., Romano, A.G. and Patterson, M.M. Extinction and retention of a classically conditioned flexor nerve response in acute spinal cat. *Behavioral Neuroscience*, 1983; 97:530-540.

Carew, T.J., Pinsker, H.M. and Kandel, E.R. Long-term habituation of a defensive withdrawal reflex in *Aplysia californica*. *Behavioral Biology*, 1972; 12:339-345.

Deese, J.E. and Kellogg, W.N. Some new data on 'spinal conditioning.' *Journal of Comparative Physiological Psychology*, 1949; 42:157-60.

Durkovic, R.G. Classical conditioning, sensitization, and habituation of the flexion reflex of the spinal cat. *Physiological Behavior*, 1975; 14:297-304.

Durkovic, R.G. and Prokowich, L.J. D-2-Amino-5-phosphonovalerate, an NMDA receptor antagonist, blocks induction of associative long-term potentiation of the flexion reflex in spinal cat. *Neuroscience Letters*, 1998; 257:162-164.

Farel, P.B., Glanzman, D.L., and Thompson, R.F. Habituation of a monosynaptic response in the vertebrate central nervous system: Lateral column-motoneuron pathway in isolated frog spinal cord. *Journal of Neurophysiology*, 1973; 36:1117-1130.

Farel, P.B. and Thompson, R.F. Habituation of a monosynaptic response in frog spinal cord: Evidence for a presynaptic mechanism. *Journal of Neurophysiology*, 1976; 39:661-666.

Fitzgerald, L.A. and Thompson, R.F. Classical conditioning of the hindlimb flexion reflex in the acute spinal cat. *Psychonomic Science*, 1967; 8:213-214.

Groves, P.M. and Thompson, R.F. Habituation: A dual-process theory. *Psychological Review*, 1970; 77:419-450.

Joynes, R.L. and Grau, J.W. Mechanisms of Pavlovian conditioning: Role of protection from habituation in spinal conditioning. *Behavioral Neuroscience*, 1996; 110:1375-1387.

Kandel, E.R. *Behavioral Biology of Aplysia*. W.H. Freeman: San Francisco, 1979.

Kellogg, W.N. Is "spinal conditioning" conditioning? *Journal of Experimental Psychology*, 1947; 37:263-265.

Kellogg, W.N., Deese, J., and Pronko, N.H. On the behavior of the lumbospinal dog. *Journal of Experimental Psychology*, 1946; 36:503-511.

Kim, J.J. and Thompson, R.F. Cerebellar circuits and synaptic mechanisms involved in classical eyeblink conditioning. *Trends in Neurosciences*, 1997; 20:177-181.

Morgan, C.T. and Stellar, E. *Physiological Psychology*. McGraw-Hill: New York, 1950.

Patterson, M.M. Mechanisms of classical conditioning and fixation in spinal mammals. In Offprints from *Advances in Psychobiology*, 1976; 3:381-434. Riesen, A.H., and Thompson, R.F. (Eds). John Wiley & Sons, Inc.

Patterson, M.M. Effects of forward and backward classical conditioning procedures on a spinal cat hindlimb flexor nerve response. *Physiological Psychology*, 1975; 3:86-91.

Patterson, M.M., Cegavske, C.F. and Thompson, R.F. Effects of classical conditioning paradigm on hindlimb flexor nerve response in immobilized spinal cat. *Journal of Comparative and Physiological Psychology*, 1973; 84:88-97.

Pinto, R. and Bromiley, R.B., A search for 'spinal conditioning' and for evidence that it can become a reflex, *Journal of Experimental Psychology*, 1950; 40:121-130.

Polenchar, B.E., Romano, A.G., Steinmetz, J.E. and Patterson, M.M. Effects of US parameters on classical conditioning of cat hindlimb flexion. *Animal Learning and Behavior*, 1984; 12:69-72.

Sherrington, C.S. *The Integrative Action of the Nervous System*. New Haven, Yale University Press, 1906.

Shurrager, P.S. and Culler, E. Conditioning in the spinal dog. *Journal of Experimental Psychology,* 1940; 26:133-159.

Shurrager, P.S. and Culler, E. Conditioned extinction of a reflex in a spinal dog. *Journal of Experimental Psychology*, 1941; 28:287-303.

Thompson, R.F. *Foundations of Physiological Psychology*. New York, Harper & Row, 1967.

Thompson, R.F. and Spencer, W.A. Habituation: A model phenomenon for the study of neuronal substrates of behavior. *Psychological Review*, 1966; 173:16-43.

Thompson, R.F. and Krupa, D.J. Organization of memory traces in the mammalian brain. *Annual Review of Neuroscience*, 1994; 17:519-549.

2 PAVLOVIAN AND INSTRUMENTAL CONDITIONING WITHIN THE SPINAL CORD: METHODOLOGICAL ISSUES

James W. Grau and Robin L. Joynes
Department of Psychology
Texas A and M University
College Station, Texas 77843

2.1 INTRODUCTION

The contributors to this volume have uncovered many interesting examples of spinal cord plasticity, from the induction of central sensitization to the recovery of function and the modification of the stepping reflex (e.g., Coderre, 1994; Durkovic, 1975; Edgerton, Roy, de Leon, Tillakaratne, and Hodgson, 1997; Patterson, Cegavske, and Thompson, 1973; Woolf, 1983). This plasticity can be studied from a number of perspectives, views that govern the kinds of questions we ask about the underlying mechanism. These perspectives differ in their relative emphasis on behavior, methodology, function, neurophysiology and neurochemistry. Research within the domain of neuroscience has emphasized the latter two components, yielding a wealth of information on the underlying biological mechanisms. Our training is in the area of learning theory and places greater weight on the former issues: on the behavioral methodologies that can be used to characterize the functional/operational properties of the system. This perspective follows in the tradition of Pavlov (1927), Hull (1943), Konorski (1967) and Rescorla (1988), building a literature that spans a century. The field of learning has provided the foundation for characterizing plasticity in other model systems (e.g., Alkon, 1984; Kandel and Schwartz, 1982). In a similar vein, we attempt to characterize the kinds of mechanisms that underlie some simple examples of spinal cord plasticity. Our work illustrates how these mechanisms can be distinguished and suggests a casual theory of how they evolved. However, before we proceed, we must first address some terminological issues and specify our criteria for learning.

2.2 DEFINING THE PROBLEM AREA

Researchers working within our field typically characterize their domain of study as learning rather than plasticity (Bower and Hilgard, 1981; Domjan, 1993; Mackintosh, 1974; Rescorla, 1988). How does learning differ from plasticity? We would like to suggest that plasticity reflects a broader term that includes neural modifications resulting from development, regeneration, and/or experience. Most reserve the term learning for the last form of plasticity, that which depends on the organism's experience. Functional changes due to development or regeneration, that do not depend on experience, represent examples of plasticity, but they do not count as learning. It is important to realize, though, that these distinctions likely reflect a fuzzy boundary. In many cases the development (or regeneration) of a system will be influenced by experience and, at the extreme, some experience may be essential. To the extent experience does matter, learning may be involved. A caveat is added to the last sentence for we need some additional criteria (see Table 2.1).

Table 2.1. Criteria for learning

1.	The modification depends on the organism's experiential history.
2.	The behavioral modification depends on a form of neural plasticity.
3.	(a) The modification outlasts (extends beyond) the environmental contingencies used to induce it. (b) The experience has a lasting effect on performance.

So far, we have adopted a definition of learning that has two basic conditions: 1) the modification depends on the organism's experiential history; and 2) the modification depends on a form of neural plasticity. To this list, most would add a third criteria (3a): the modification outlasts (extends beyond) the environmental contingencies used to induce it. For example, a rat will normally orient towards an auditory source (e.g., a tone). Presenting another stimulus (e.g., a light) would decrease the likelihood of this response. But suppose that as soon as we stop presenting the light, subjects again orient to the tone. In this case, we would say that the change in the orienting response reflects a *performance effect* rather than learning. A corollary to criterion 3 is that the neural modification has a lasting effect upon behavior (3b). Of course, this raises the thorny issue of defining "lasting" to which no one has provided a definitive answer. In practice, though, it appears that most require that the modification last minutes, hours, or days, rather than seconds. Notice that 3b implies a form of memory (information storage). As Domjan (2000) notes, learning and memory are like two sides of a coin, you cannot have one without the other.

Because our definition of learning hinges on the term *experience*, we need to be clear about what this term entails. We adopt the common view that the organism's experiential history includes environmental stimulation and behavior that lie within the normal operational range of the system. Such a definition helps to exclude problematic cases that reflect dysfunction (e.g., neural insults) rather than learning. It also excludes behavioral modifications resulting from physiological manipulations (e.g., neural regeneration, infusion of growth factor). Of course, physiological manipulations can be used to study learning, but this work must be placed in a larger theoretical perspective that relates the process to the mechanisms that the organism normally uses to encode its experiential history.

At this juncture, how one proceeds depends on one's theoretical orientation and thus, we must be up front about our own. Our primary concern is with the functional mechanisms that underlie learning and we assume that a variety of mechanisms (both associative and nonassociative) normally contribute (Rescorla, 1988). The emphasis on mechanism distinguishes our approach from the descriptive approach of the Skinnerians (Skinner, 1938) who avoid theorizing about internal mechanisms and who focus instead on the environmental contingencies. The assumption that a variety of mechanisms underlie learning contrasts with earlier perspectives that focused on associative processes alone, taking pains to tease out the contaminating nonassociative components (Gormezano and Kehoe, 1975; Kimble, 1961). From our perspective, a variety of mechanisms may encode environmental relations and there is little reason to believe, a priori, that one is superior to the rest.

To study the way in which organisms encode environmental relations, researchers have typically used three basic methodologies: single stimulus learning, Pavlovian (or classical) conditioning, and instrumental conditioning (Domjan, 1993; Rescorla, 1988). Single stimulus learning represents the simplest case. Here, the organism experiences (nominally) just one stimulus (S). In many cases, presenting a stimulus alone undermines its behavioral consequences, a phenomenon known as habituation. At other times, prior exposure to a stimulus may enhance its behavioral impact, an effect known as sensitization. Pavlovian conditioning occurs when the response elicited by one stimulus (S1) depends on its temporal relationship with another stimulus (S2). Normally, a biologically significant event that unconditionally elicits a response serves as S2. Consequently, S2 is often referred to as an unconditioned stimulus (US) and the behavior elicited is called the unconditioned response (UR). The stimulus we are attempting to train (S1) is generally called the conditioned stimulus (CS) and the behavior it elicits is termed the conditioned response (CR). The classic example from Pavlov's laboratory involves pairing a tone (the CS) with food (the US). After a number of tone-food pairings, dogs begin to exhibit a conditioned salivation response (the CR) to the tone. The third common category of learning depends on the temporal relationship between a response (R) and a stimulus outcome (O). The outcome is sometimes called the "reinforcer" and generally involves an appetitive (e.g., food) or aversive (e.g., shock) event. For example, a hungry rat will learn to press a bar (the R) to obtain food (the outcome). Researchers (Hilgard and Marquis, 1940; Mackintosh, 1974; Rescorla, 1988) that follow in the tradition of Thorndike (1898) and Hull (1943) often refer to this kind of learning as "instrumental" whereas Skinnerians prefer the term "operant" (Skinner, 1938).

Our goal is to use these methods to characterize the functional principles that govern some simple examples of spinal cord plasticity. The primary question is: should these examples of plasticity be considered instances of learning, and if so, what kind of mechanism is involved? This goal, combined with our desire to relate the work to studies of learning in intact subjects, placed some special demands on our experimental paradigm. For example, to institute the environmental relations needed to infer mechanism, and to address methodological problems that have plagued past studies, multiple stimuli and responses were required along with a host of controls. Adopting these procedures allowed us to collect the data needed to convince skeptics (and there were many) that spinal cord systems are capable of learning. But convincing the skeptics required that we play by their rules and to do so we needed multiple experimental controls embedded in large factorial designs. Experiments of this kind typically require 32 subjects and a complex 3 x 4 factorial design may need 72 subjects or more. And dozens of experiments needed to be done. To collect this data in a timely fashion, we had to simplify some aspects of our experimental procedure. This was accomplished by using stimuli (transcutaneous electrical stimulation, radiant heat) and responses (tail-flick, leg flexion) that could be easily applied and monitored. The benefit of this procedure was that it generated data that could be readily compared to that obtained using intact subjects and it allowed us to run the myriad of controls needed to uncover the principles of learning. Thus, our strategy was to first characterize the underlying principles and convince our colleagues that the spinal cord does support some interesting forms of learning. Once this was accomplished, we hope to proceed with a more detailed analysis of the underlying physiological systems. In the meantime, we must acknowledge that we know relatively little about the underlying biological mechanisms, though some hints will be provided.

In the next section of the paper, we will briefly review our studies on Pavlovian conditioning (Grau, Salinas, Illich, and Meagher, 1990; Illich, Salinas, and Grau, 1994; Joynes and Grau, 1996; Joynes, Illich, and Grau, 1997) and show how, given some basic behavioral operations, we can gain insights into the nature of the underlying mechanism. Next, we consider our studies on instrumental learning (Grau, Barstow, and Joynes, 1998; Joynes, Janjua, and Grau, submitted; Joynes,

Crown, Ferguson, and Grau, submitted). Though early studies suggested that spinal cord neurons are capable of this form of learning (Buerger and Chopin, 1976; Buerger and Fennessy, 1970; Chopin and Buerger, 1976; Horridge, 1962), this conclusion has been questioned on methodological grounds (Church, and Lerner, 1976). We show how these methodological issues can be addressed to provide unequivocal evidence that spinal cord systems are sensitive to response-outcome relations.

2.3 PAVLOVIAN CONDITIONING

Our first studies were motivated by a theoretical claim regarding the way in which learning affects pain (Grau, 1987). Considerable evidence indicates that pain modulatory systems are sensitive to Pavlovian relations (Fanselow, 1986). This has been shown by pairing a cue (the CS) with an aversive shock (the US). After just a few CS-US pairings, the CS acquires the capacity to inhibit nociceptive responses (e.g., tail withdrawal from radiant heat [the *tail-flick test*]). This effect (see Figure 2.1A) is known as *conditioned antinociception* and we (Grau, 1987) and others (Maier, 1989; Watkins and Mayer, 1982) suggested that this learning depends on forebrain systems.

At the same time, it was known that intraspinal mechanisms can generate antinociception (Meagher, Chen, Salinas, and Grau, 1993). In these studies, Mary Meagher transected the spinal cord at the second thoracic vertebra (T2). After a recovery period of 8-10 hrs, rats were exposed to a few intense (3 mA) tailshocks. Tailshock caused a dramatic increase in tail-flick latencies suggesting that afferent nociceptive signals can, in some circumstances, directly engage intraspinal mechanisms that inhibit nociceptive reactivity. We assumed that this antinociceptive system was not sensitive to learning--that it reflected a passive, unconditioned, response to noxious stimulation.

An undergraduate in our laboratory, Juan Salinas, challenged this theoretical claim. Noting that Patterson, Thompson, Durkovic and others had found evidence that spinal cord systems could support Pavlovian conditioning (Beggs, Steinmetz, and Patterson, 1985; Durkovic, 1975; Fitzgerald and Thompson, 1973; Shurrager and Culler, 1940), he argued that antinociceptive mechanisms within the spinal cord might be sensitive to environmental relations. Given that our own theory suggested otherwise (Grau, 1987), we did not immediately embrace his proposal. Nonetheless, he persisted, leading us to develop a procedure that might provide a reasonable test of the hypothesis (Figure 2.1B). Following Kandel and his colleagues (Kandel and Schwartz, 1982), we used a differential conditioning paradigm in which one stimulus (the CS+) was paired with the US while another was presented alone (the CS-). At the end of training, the magnitude of the response elicited by the CS+ was compared to that generated by the CS-. The advantage of this paradigm is that it controls for the consequences of exposure to the CS and US alone (pseudoconditioning and sensitization). Under these conditions, evidence that the CS+ and CS- differentially affect behavior suggests that the system must be sensitive to the CS-US relation.

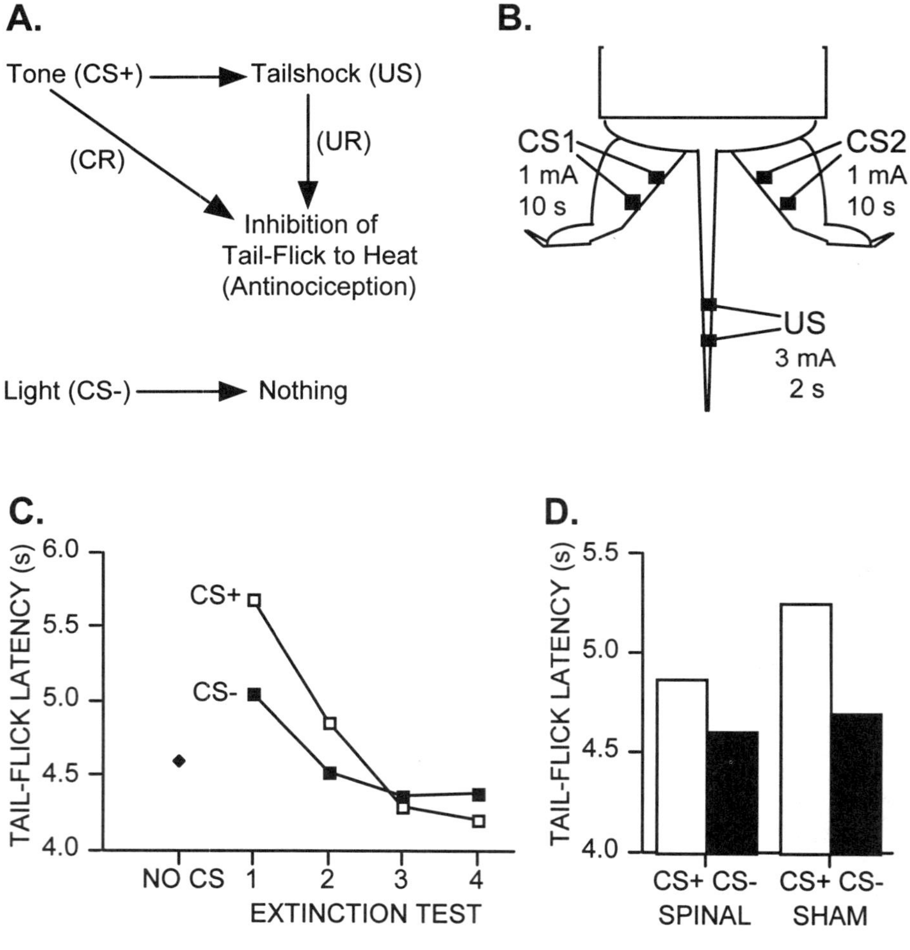

Figure 2.1. (A) The relations that underlie conditioned antinociception in intact rats. (B) The stimuli used to study conditioned antinociception in spinally transected rats. (C) A CS that has been paired with the US (CS+) generates longer tail-flick latencies in spinal rats than a CS that was presented alone (CS). This difference extinguishes over the course of testing (blocks of 2 trials). (D) Impact of a CS+ and CS- on rats that were trained with the spinal cord intact and then received either a spinal transection or a sham operation.

Procedure

To study whether spinal antinociceptive systems would exhibit differential conditioning, rats received a spinal transection at T2 (we will refer to such subjects as *spinal,* or *spinalized,* and in all cases the transection was at T2). Ten (early studies) to 24 hrs (later studies) after surgery, subjects were placed in opaque tubes designed to comfortably hold the subject without inducing undue stress (i.e., "loose" rather than "tight" restraint). For our CSs we used moderate shocks (10-s, 1.0-mA) that we thought, on the basis of past work (Meagher et al., 1993), would generate little antinociception. These were applied through Michel clips attached to each leg as illustrated in Figure 2.1B. For our US, we used a 2-s, 3-mA, shock that was presented through cutaneous tail electrodes. Prior research had established that tailshocks at this intensity induced a strong unconditioned antinociception in spinalized rats (Meagher et al., 1993). Stimulation to one hind leg (the CS+) was paired with tailshock using a delayed conditioning procedure in which the US occurred during the last 2 seconds of the CS+. Stimulation of the other hind leg served as the CS- which was presented alone (during the intertrial interval [ITI]) an equal number of times. Training consisted of 30 presentations of the CS+ (CS-US pairings) and CS- given in a random order approximately 60 s apart. Subjects were then given an hour break to allow any unconditioned antinociceptive effects of the US to dissipate. After this rest period, subjects were returned to the restraining tubes and baseline nociceptive thresholds were assessed using the tail-flick test. Finally, tail-flick latencies during the CS+ and CS- were tested in a counterbalanced order 8 times each.

In this experiment, and all subsequent studies, statistical significance was evaluated using an analysis of variance (ANOVA) followed by appropriate post hoc comparisons (Duncan's new multiple range, Newman-Keuls, or Bonferroni *t*, all of which are designed to maintain the criterion for statistical significance [alpha] at .05 for a family of contrasts). In some of our more complex experimental designs, trend analysis was used to evaluate the significance of the linear, quadratic (one inflection), cubic (two inflections), or quartic (three inflections) components of a trials or drug effect. Unless otherwise noted, all of the described group differences were statistically significant ($p < .05$). Furthermore, most accounted for a very large proportion of the variance, yielding eta squared values (proportion of variance accounted for) in excess of 40% and effects sizes ($d \geq 1.2$) well above the traditional definition of "large" ($d = 0.8$) (Cohen, 1988).

Evidence for Conditioned Antinociception in Spinalized Rats

The impact of the CS+ and CS- on tail-flick latencies after differential conditioning is illustrated to the right of the baseline scores in Figure 2.1C. It is evident that the CS+ induced greater antinociception (longer tail-flick latencies) and that this effect dissipated (extinguished) over the course of testing.

How do these CS+/CS- differences compare to those typically observed in intact subjects? An answer to this question was provided by an experiment designed to assess whether these spinally mediated changes in nociceptive plasticity develop in intact subjects. This is an important issue for, in other preparations (e.g., Satinoff, 1978), brain systems sometimes usurp control over lower-level mechanisms within the spinal cord. Given this, it is possible that conditioned antinociception in intact subjects always depends on brain-mediated processes. If so, conditioned antinociception should be eliminated by cutting the spinal cord between training and testing. To examine whether this occurs, we trained intact rats using cutaneous CSs applied to the plantar surface of the left or right hindpaw and a tailshock US (Joynes et al., 1997). After training, half the subjects received a spinal cord transection while the remaining rats underwent a sham operation. Tail-flick latencies during the CSs were assessed the next day. Note that the design of this experiment worked against observing a CS+/CS- difference in spinalized rats, for less intense stimuli were used during training and the retention interval was increased from 1 to 24 hrs. Nonetheless, all but 1 of the 12 spinalized rats exhibited longer tail-flick latencies during the CS+, a difference that accounted for 40% of the variance (Figure 2.1D). Sham operated rats also exhibited conditioned antinociception and the magnitude of this effect appeared somewhat larger (roughly, 2 fold), implying that brain-mediated processes normally amplify the CR.

Mechanisms of Pavlovian Conditioning

Our results verify that spinal cord systems are sensitive to the CS-US relations that underlie Pavlovian conditioning. It must be remembered, however, that Pavlovian conditioning refers to a methodology and that a variety of mechanisms can potentially encode the S1-S2 relation. Three of these mechanisms are depicted in Figure 2.2. At the top, we have illustrated *associative learning* under idealized conditions. In this case, the CSs do not generate a CR-like response prior to training. When one CS (+) is paired with the US, it acquires the ability to generate a CR while the unpaired CS (-) does not. At the end of this differential conditioning, the impact of the CSs is compared to the effect of a novel CS (N). As illustrated in the histogram, only the CS+ generates a CR.

The middle set of graphs illustrate an alternative mechanism, *pairing specific enhanced sensitization*. This mechanism, which is also known as activity-dependent presynaptic facilitation (Hawkins, Abrams, Carew, and Kandel, 1983) or activity-dependent neuromodulation (Walters and Byrne, 1983), is thought to underlie Pavlovian conditioning in the invertebrate *Aplysia*. In this case, the CSs elicit a CR-like response prior to training and exposure to the US alone can further sensitize this pre-existing response. However, the magnitude of this process is thought to depend on the CS-US relationship; greater sensitization develops when the CS is paired with the US. As a consequence, the CS+ generates a stronger conditioned response at the end of testing relative to both the CS- and a novel CS.

The last mechanism, *protection from habituation*, has received somewhat less attention. Again, it is assumed that the CSs generate a CR-like response at the start of training. It is well established that presenting the CS alone can cause such a response to decline (habituate). It is also known that habituation can be disrupted by the presentation of an extraneous event (dishabituation), such as a US. If the magnitude of this effect varies as a function of the CS-US interval, being greatest when the two events are paired, presentation of the US could produce a CS+/CS- difference by protecting the CS from habituation (Humphrey, 1933; Mitchell, Scott, and Mitchell, 1977; Pfautz, Donegan, and Wagner, 1978; Rescorla, 1984; Riley, 1978; Whitlow and Wagner, 1984). But in this case, the CS+ would not generate a response greater than that elicited by a novel CS.

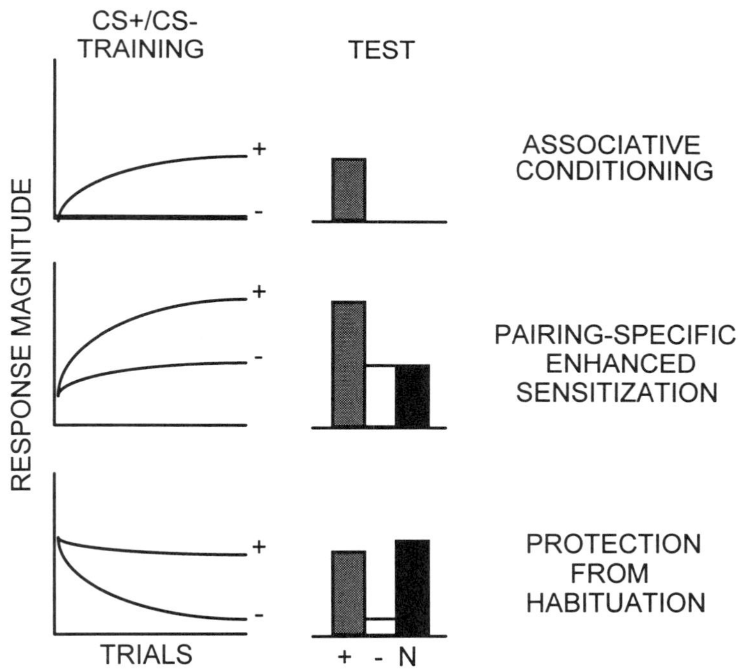

Figure 2.2. Three ways in which instituting a CS-US relationship can influence reactivity to the CS. The left-most panels depict reactivity to the CS+ and CS- during differential conditioning under idealized conditions. The histograms positioned to the right of these acquisition curves illustrate the expected levels of performance to the CS+, the CS- and a novel CS (N) at the end of training. (Adapted from Joynes and Grau, 1996, *Behavioral Neuroscience*, *110*, 1375-1387. © 1996 American Psychological Association.)

There is a tradition within psychology that the only "true" form of learning involves associative learning and that this mechanism is the only one worthy of study (Gormezano and Kehoe, 1975; Kimble, 1961). From this perspective, much turns on the issue of preexisting responses. If the CS elicits a miniature copy of the CR prior to training, the learning is relegated to the realm of alpha conditioning and deemed not worthy of study. We believe that this tradition has hampered progress. By ignoring this stricture, Kandel, Byrne, Walters and their colleagues have made tremendous progress in elucidating the biological mechanisms involved in pairing specific enhanced sensitization and have shown how this mechanism could contribute to a variety of behavioral phenomenon (Hawkins et al., 1983; Hawkins and Kandel, 1984; Walters and Byrne, 1983). Protection from habituation may have an equally pervasive role in regulating behavior and understanding its operation would appear just as important. For example, attentional theorists often assume that pairing a CS with a US helps maintain attention to the CS, and thereby facilitates the abstraction of the CS-US relation (Lubow, Schnur, and Rifkin, 1976). This is just a protection from habituation account couched in attentional terms.

Role of Protection from Habituation

Given these distinctions, we sought to unveil the mechanism(s) that underlie(s) the development of conditioned antinociception within the spinal cord (Joynes and Grau, 1996). We reasoned that it would be best to begin with the simplest assumption, that the learning reflected a form of protection from habituation. If this was true, the CSs should elicit a CR-like response prior to training.

We had assumed that our CSs would not affect nociceptive reactivity because prior studies had shown that stimulation at this level (1-mA) has no effect on tail-flick latencies (Grau, Burks, Kallina, King, and Meagher, 1996; Meagher et al., 1993). However, these studies used cutaneous electrodes and assessed tail-flick latencies after shock. In contrast, our conditioning studies (Grau et al., 1990; Illich et al., 1994) utilized intracutaneous electrodes and assessed nociceptive reactivity while the stimulus was being presented. Given this, more direct evidence was required. This was obtained by assessing the impact of our CSs on tail-flick latencies in naïve subjects. To our surprise, we found that the CSs induced antinociception prior to training, producing an increase in tail-flick latencies comparable to that observed during the CS+ (Joynes and Grau, 1996).

Our next experiment assessed whether exposure to the CS alone would weaken the CS induced antinociception. Spinal rats experienced one CS 30 times over the course of an hour. At the end of training, we determined whether this experience caused the CS-induced antinociception to habituate by comparing the preexposed cue to a novel CS (stimulation of the contralateral leg). The results are illustrated in Figure 2.3A and reveal that the preexposed CS (CS-H) generated a weaker antinociception than the novel CS (CS-N).

Our findings suggested that the CS+/CS- difference observed in earlier studies might reflect a form of protection from habituation. To evaluate this possibility more directly, we assessed whether pairing a CS with a US (tailshock) would alter the rate at which habituation develops (Joynes and Grau, 1996). One group of spinalized rats experienced 30 CS-US pairings while another experienced the CS and the US explicitly unpaired. An hour after training, we compared the antinociceptive impact of the preexposed CS to a novel CS (Figure 2.3B). Subjects that experienced the CS and US in an unpaired fashion exhibited less antinociception during the preexposed CS (the CS-) than a novel CS (CS-N). Pairing the CS with the US (CS+) appears to have protected it from habituation, allowing it to generate an antinociception that was as strong as that produced by the novel CS.

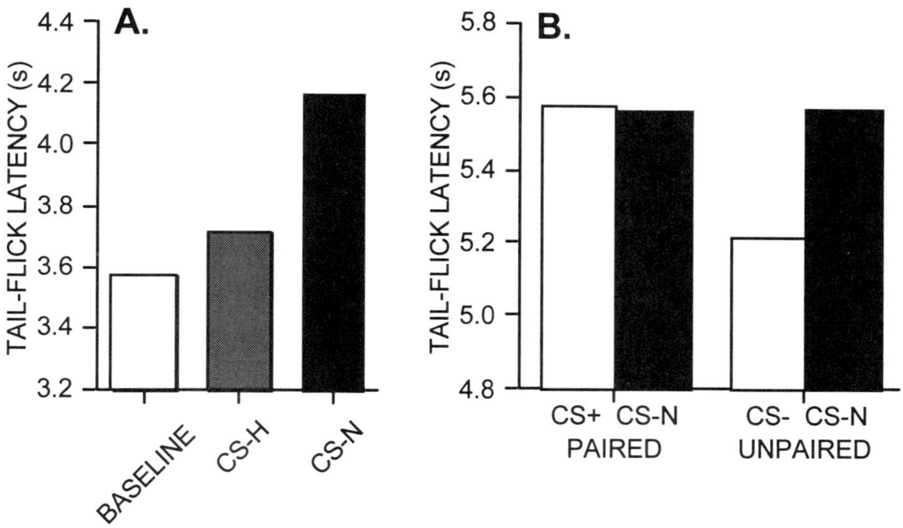

Figure 2.3. (A) Mean tail-flick latencies in the absence of a stimulus (BASELINE), during the preexposed conditioned stimulus (CS-H), and during a novel CS (CS-N). (B) Tail-flick latencies in spinal rats that received the CS paired with the US (CS+) or the CS and US explicitly unpaired (CS-). In each case, reactivity to the pretrained CS was compared to a novel CS (CS-N). (Adapted from Joynes and Grau, 1996, *Behavioral Neuroscience*, *110*, 1375-1387. © 1996 American Psychological Association.)

At this point, it is tempting to conclude that the learning reflects protection from habituation. However, there is another equally valid interpretation of the data. Spinal rats may have exhibited longer tail-flick latencies during the CS+ because associative learning effectively counteracted the habituation. To address this possibility, we examined the effect of two manipulations that should affect habituation and associative learning in opposite ways. The first involved varying the intertrial interval (ITI), the period between CS-US pairings. Considerable evidence indicates that, within limits, increasing the ITI (spaced practice) improves associative learning (Gibbon and Balsam, 1981; Miller and Matzel, 1989; Prokasy, Gran, and Myers, 1958; Spence and Norris, 1950). In contrast, nonassociative habituation develops more rapidly when trial spacing is decreased (massed practice) and little habituation is observed with long ITIs (Davis, 1970; Groves, Lee, and Thompson, 1969; Wagner, 1976; Whitlow, 1975). If our results reflected a form of protection from habituation, increasing the ITI should lessen habituation to the CS- and thereby decrease the CS+/CS- difference. In contrast, an associative account anticipated that increasing the ITI should, if anything, promote the development of a CR to the CS+ and thus augment the CS+/CS- difference. To examine these alternatives, spinal rats received the CS in either a paired (CS+) or unpaired (CS-) fashion using either our usual 2 min ITI or one that was 5 times longer (10 min). As shown in Figure 2.4, rats trained with a 2 min ITI exhibited a pattern of results identical to those obtained earlier; the unpaired CS exhibited a weaker antinociception relative to both a novel CS and one that had been paired with the US. Increasing the ITI eliminated habituation to the CS- and thereby eliminated the CS+/CS- difference, an outcome indicative of protection from habituation rather than associative learning.

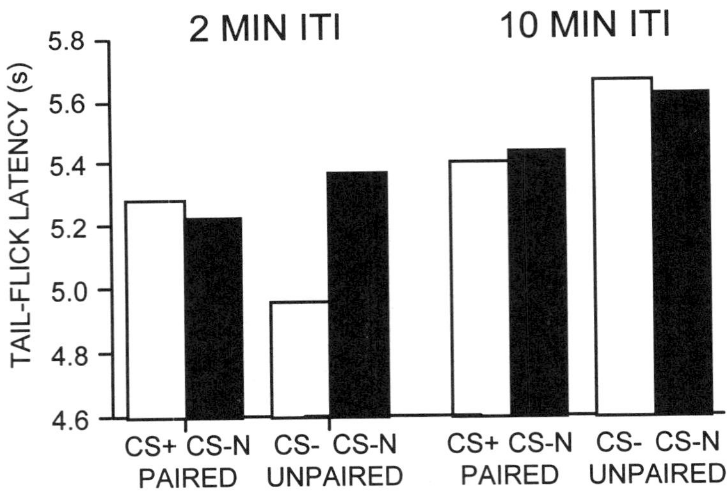

Figure 2.4. Mean tail-flick latencies in spinal rats that received the CS paired with the US (CS+) or the CS and US explicitly unpaired (CS-). The stimuli were presented on a variable intertrial interval (ITI) with a mean of either 2 (left) or 10 (right) min. In all cases, reactivity to the pretrained CS was compared to a novel CS (CS-N). (Adapted from Joynes and Grau, 1996, *Behavioral Neuroscience*, *110*, 1375-1387. © 1996 American Psychological Association.)

Our last experiment in this set examined the impact of varying the number of training trials. Again, different results were expected depending on the mechanism involved. Theories of associative learning assume that increasing the number of training trials should strengthen the CS+/CS- difference (e.g., Rescorla and Wagner, 1972). In contrast, because the protection from habituation mechanism slows, but presumably does not abolish habituation to the CS+, increasing the number of training trials could foster habituation to the CS+ and undermine the CS+/CS- difference. We evaluated these alternatives by assessing the impact of increasing the number of training trials five fold, from 30 to 150. In both cases, a 2 min ITI was used and subjects were tested an hour after the last training trial. The results are depicted in Figure 2.5. As usual, subjects given 30 exposures exhibited habituation to the CS- but not the CS+. Increasing the number of training trials caused an increase in the antinociceptive effect of the novel CS (CS-N). This likely reflected a general sensitization effect attributable to the additional USs. Of greater interest, increasing the number of training trials fostered habituation to the CS+ and this in turn eliminated the CS+/CS- difference. Again, this runs counter to an associative account and suggests that this learning reflects a form of protection from habituation.

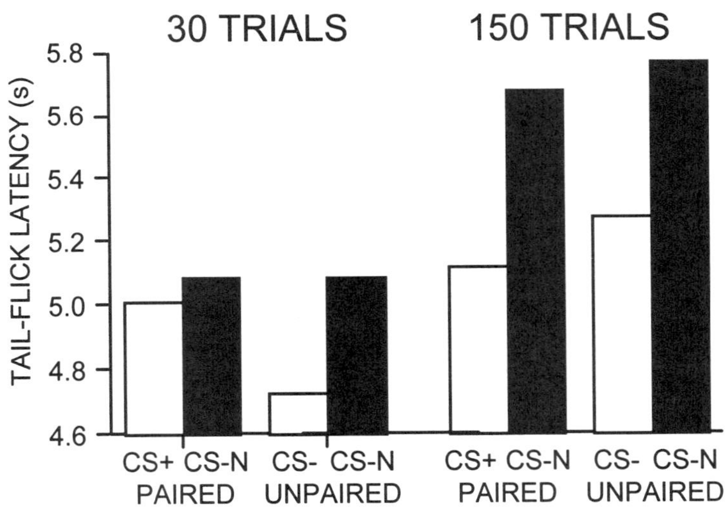

Figure 2.5. Mean tail-flick latencies in spinal rats that received either 30 (left) or 150 (right) training trials with the CS paired with the US (CS+) or the CS and US explicitly unpaired (CS-). In all cases, reactivity to the pretrained CS was compared to a novel CS (CS-N). (Adapted from Joynes and Grau, 1996, *Behavioral Neuroscience*, *110*, 1375-1387. © 1996 American Psychological Association.)

Implications and Limits

Having elucidated the mechanism that underlies spinally-mediated conditioned antinociception, two cautions are warranted. First, it would be a mistake to conclude that all examples of spinally-mediated Pavlovian conditioning necessarily reflect protection from habituation. This is an empirical issue that requires further research. To discount this possibility, researchers must show that the CS+ produces a CR greater than that produced by a novel CS and that increasing either the ITI or number of training trials enhances the CS+/CS- difference. The second caution concerns a common preconception: that protection from habituation reflects a simple process that does not support sophisticated Pavlovian phenomena. Against this preconception, we have shown that: 1) the CS+/CS- difference declines over the course of testing (extinction); 2) pre-exposure to the CS alone undermines learning (latent inhibition); and 3) concurrent presentation of a more salient (intense)cue can prevent learning (overshadowing) (Grau et al., 1990; Illich et al., 1994). Others have suggested that exposure to the CS+ alone, either before or after conditioning, induces a form of habituation that undermines its effectiveness. This presumably accounts for the latent inhibition and extinction effects observed in our spinal paradigm. Overshadowing could reflect a form of crossed inhibition (Sherrington, 1906) whereby a more salient CS inhibits the processing of less intense stimuli applied to the contralateral leg. Irrespective of how these phenomenon occur, they demonstrate that even the simplest mechanism of Pavlovian conditioning can exhibit a surprising range of Pavlovian phenomena. Indeed, it is tempting to speculate that these phenomena are inherent properties of neural systems capable of Pavlovian conditioning. At the same time, it is important to acknowledge that there are a variety of more sophisticated behavioral effects that we suspect lie outside the limits of spinal systems (see Table 2.2). These include configural learning (e.g., negative patterning; Sutherland and Rudy, 1989), sensory preconditioning, trace conditioning (Solomon, Vander Schaaf, Thompson, and Weisz, 1986), a variety of performance effects (Miller and Matzel, 1989), and the mediated acquisition and extinction effects described by Holland (1990). Evidence suggests that such phenomena depend on neural systems in the brain (e.g., hippocampus) and may reflect (require) a form of associative learning (Squire and Zola, 1996; Sutherland and Rudy, 1989).

Table 2.2. Learning mechanisms: Functional properties

Protection from Habituation	Pairing-Specific Enhanced Sensitization	Associative Conditioning
Differential Cond.		Negative Patterning
Extinction	?	Sensory Preconditioning
Latent Inhibition		Mediated Acquisition
Cue Competition		Mediated Extinction

We have contrasted protection from habituation to associative learning, remaining curiously silent regarding the limits and capacities of pairing specific enhanced sensitization. We have adopted this course for a simple reason: there is currently not enough data to constrain theorizing. Moreover, it may be unreasonable to expect a simple answer, for we may be dealing with a continuum of learning potentials rather than three qualitatively distinct systems. At the extreme, a US might only sensitize a CS-elicited behavior when the two events are paired. In this case, the learning could appear associative in nature and abide by its rules. In other cases, the mechanism could be less well developed and yield a pattern of results more akin to protection from habituation.

Notice that this perspective suggests a casual theory of how conditioning mechanisms may have emerged. At the simplest (pre-Pavlovian) level, whether or not the dishabituating or sensitizing stimulus is paired with the target stimulus could be irrelevant. An elaboration of the dishabituation process, one that supports protection from habituation, might have then produced the simplest form of conditioning. Quantitative enhancement of this process could yield a mechanism that not only protects a response from habituation but selectively augments it. The last step, the emergence of associative learning, may depend on the development of a new process, one that supports the perception of environmental events and allows a CS to evoke an image, or expectation, of the US (Holland, 1990).

2.4 INSTRUMENTAL CONDITIONING

We now turn our attention to a different form of learning, instrumental conditioning. Earlier studies had presented evidence that spinal cord systems are sensitive to response-outcome relations (Buerger and Chopin, 1976; Buerger and Fennessy, 1970; Chopin and Buerger, 1976), the defining property of instrumental learning (Bower and Hildgard, 1981; Domjan, 1993, Mackintosh, 1974). However, these experiments have been challenged on methodological grounds (Church, 1989; Church and Lerner, 1976) and, consequently, current dogma holds that instrumental learning in vertebrates requires a brain. Given this state of affairs, we set out to reexamine the issue using methods designed to address the criticisms of past studies (Grau, Barstow, and Joynes, 1998).

Criteria for Instrumental Conditioning

Our first task was to derive a set of criteria for demonstrating instrumental learning. These are listed in Table 2.3. We will begin by focusing on a set of minimum criteria, those common to most definitions of instrumental learning (Bower and Hildgard, 1981; Domjan, 1993; Hearst, 1975; Mackintosh, 1974; Rescorla, 1988). After we review our findings, we will consider the more advanced criteria that may serve to distinguish complex forms of "voluntary behavior" from simpler forms of response-outcome learning.

Table 2.3. Criteria for instrumental and operant learning

Minimum Criteria (Instrumental)
1. Instituting a relationship between the response and an outcome produces a change in behavior (performance).
2. The effect is neurally mediated.
3. The modification outlasts (extends beyond) the environmental contingencies used to induce it.
4. The behavioral modification depends on the temporal relationship between the response and the outcome.
Advanced Criteria (Operant)
5. The nature of the behavioral change is not constrained (e.g., either an increase or decrease in the response can be established).
6. The nature of the reinforcer is not constrained (a variety of outcomes can be used to produce the behavioral effect).

The first three criteria are directly related to the conditions listed in Table 2.1. They differ in that Criterion 1 specifies the common condition for instrumental learning: Instituting a relationship between a response and an outcome produces a change in behavior (Domjan, 1993; Bower and Hilgard, 1981; Mackintosh, 1974; Rescorla, 1988). The next two criteria specify general requirements for learning. These require evidence that the behavioral modification: is neurally mediated (e.g., it does not reflect an exercise or fatigue effect attributable to a change in the muscles that mediates the response); and outlasts (extends beyond) the environmental contingencies used to induce it. The last criterion seeks evidence that the response-outcome relationship is encoded. The problem is that a reactive model, a mechanical system incapable of learning, can potentially generate performance differences in the absence of learning. As we will see, addressing this issue represents the greatest challenge to demonstrating that a system is capable of instrumental learning. We will attempt to discount this possibility by demonstrating that the contingencies have a lasting effect on behavior (Criterion 3) and by providing evidence that learning depends on the temporal relationship between the response and the outcome (Criterion 4).

Procedure

Researchers have traditionally used a variation of the Horridge procedure to study whether spinal cord neurons are sensitive to response-outcome relations. The apparatus used in our laboratory is illustrated in Figure 2.6A. Legshock is applied through electrodes inserted into the tibialis anterior muscle. Shock elicits a leg flexion and leg position is monitored by means of a contact electrode (a metal rod) attached to the ventral surface of the paw. Whenever the rod contacts the underlying salt solution, it completes a circuit. This provides a measure of leg position (the response) that can be used to institute a response-outcome contingency. The consequences of this contingency can be monitored by measuring three variables (see Figure 2.6B); flexion number (incremented each time the contact electrode leaves the solution), time in the solution, and flexion duration. To evaluate how the response changes as a function of training duration, the session is broken into a series of 60 s time bins and for each, the mean number of flexions and net time in solution are recorded. Flexion duration is then derived using the following formula: flexion duration = time out of solution/(flexion number + 1). Given a bin duration of 60 s, time out of solution is specified by: 60 s - time in solution (in seconds).

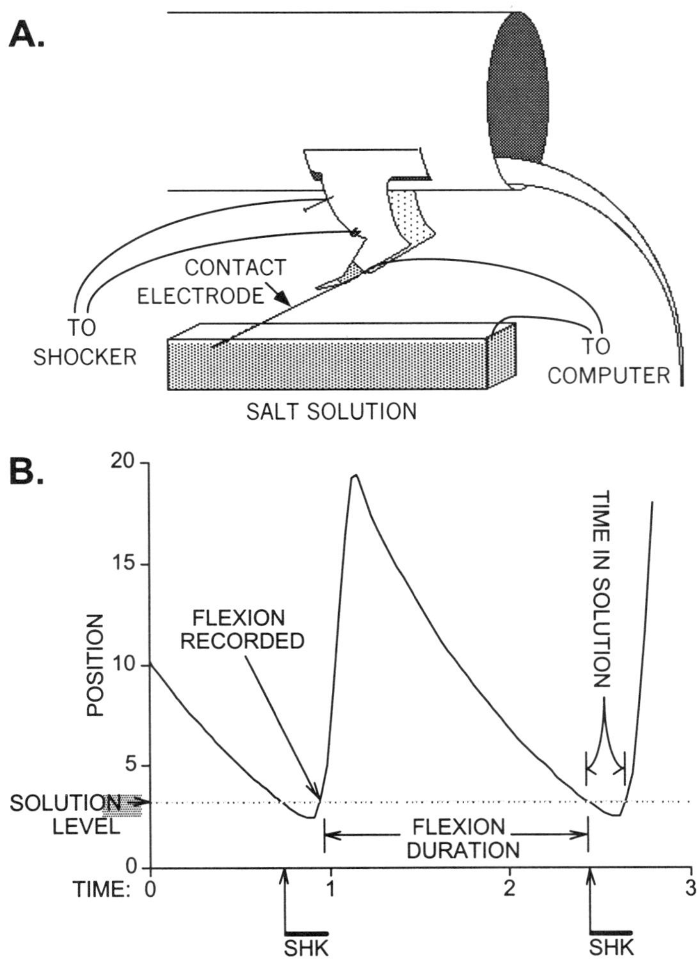

Figure 2.6. (A) The apparatus used to apply leg shock and monitor leg position in spinal rats. (B) The behavioral parameters monitored. The function depicts a snapshot of behavior during the course of training and represents the location (in millimeters) of the contact electrode tip (relative to the resting position) as a function of time (in seconds). (Adapted from Grau et al., 1998, *Behavioral Neuroscience*, *112*, 1366-1386. © American Psychological Association.)

In a typical study, one group of rats (master) receives legshock whenever the rod contacts the solution. If an intact rat were exposed to this contingency, it would quickly learn to avoid shock by maintaining its leg above the solution. Behaviorally, this would produce a decrease in time in solution, flexion number, and an increase in flexion duration. That an intact subject does this is hardly surprising. What is surprising is that this contingency can also produce a systematic change in behavior after the spinal cord has been transected. A critic could charge, however, that this behavioral modification has nothing to do with instrumental learning, but instead reflects a form of single stimulus learning in which shock *per se* sensitizes the response (Woolf, 1983, 1984). To address this possibility, researchers have typically included a yoked control group. Each subject in the yoked group is coupled to a master rat and receives legshock whenever the master subject receives a shock. Thus, subjects in the yoked group receive the same amount of shock but it is presented in a noncontingent fashion, irrespective of leg position. If all that matters is the net amount of shock experienced, not the response-outcome relation, yoked rats should also exhibit an increase in flexion duration. If they do not, we have one indication that the behavior observed in the master rats reflects a form of instrumental learning (see the bottom panel of Figure 2.7).

Criticisms of the Master-Yoke Paradigm

Unfortunately, we cannot rely on the master-yoke paradigm alone, for it can generate behavioral differences in the absence of instrumental learning (Church, 1964; Church and Lerner, 1976. For example, suppose that shock always elicits an equally strong flexion response, but the rate at which the leg falls varies in a probabilistic fashion, as illustrated in the top panel of Figure 2.7. Given this simple reactive model, the yoked rat should contact the solution first on roughly half the trials. When this occurs, it will continue to contact the solution until the master rat contacts the solution and both rats receive shock. In contrast, the master rat will always receive shock as soon as the contact electrode touches the solution and, as a consequence, will spend less time in the solution. Notice too that the master rat should, on roughly half the trials, contact the solution first. On these trials, the yoked rat does not earn a response for its contact electrode has not reached the solution. Given this, a reactive model predicts that master rats will exhibit roughly twice as many responses, a behavioral pattern observed in some prior studies (Church and Lerner, 1976; Eisenstein, 1970). Conversely, because yoked rats respond at a lower rate, and because flexion duration is inversely related to response number, the reactive model predicts that yoked rats would have longer flexion durations than master subjects. This is exactly the opposite of what we would expect given a system capable of instrumental learning.

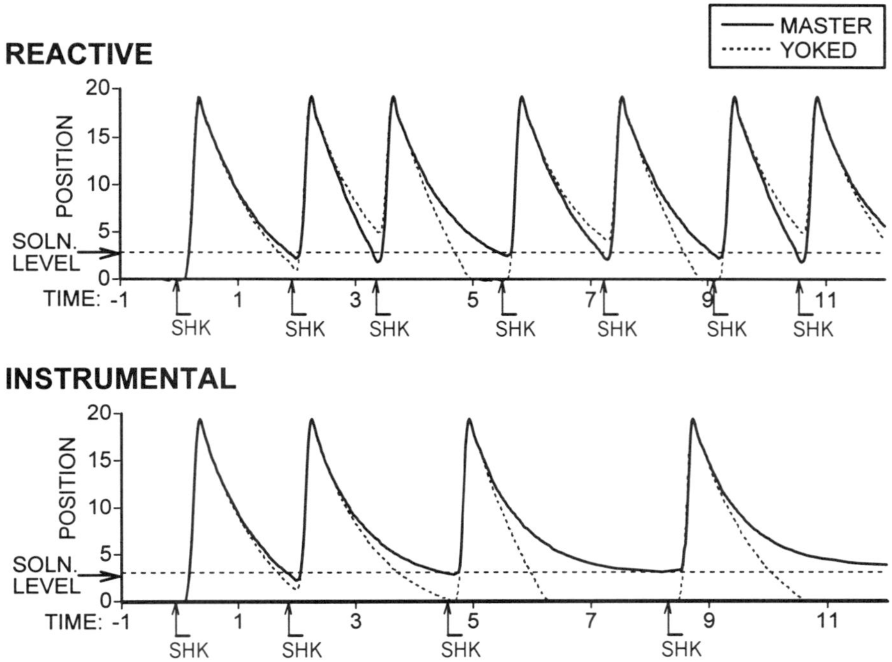

Figure 2.7. Simulated output for a reactive model (top panel) and a system capable of instrumental learning (bottom panel) in the Horridge paradigm. Each graph depicts the location of the contact electrode (in millimeters above the resting position) as a function of time (in seconds). The solid line indicates the performance of a master subject that receives contingent shock (SHK), and the dashed line represents the behavior of a yoked rat that receives shock irrespective of leg position. The onset and duration of the shocks are indicated below the *x*-axis. In this example, statistical variation in fall rate yielded the performance differences observed with the reactive model (see Grau et al. [1998] for further details). Instrumental learning was simulated by adding one assumption; that response contingent shock causes a decrease in fall rate. (Adapted from Grau et al., 1998, *Behavioral Neuroscience*, *112*, 1366-1386. © American Psychological Association.)

How can we address these problems and reject the reactive model? First, our primary measure of learning needs to be flexion duration rather than time in solution. Whereas both the reactive and instrumental model anticipate that master rats will contact the solution less, only the instrumental account predicts an increase in flexion duration. Second, we need to evaluate whether exposure to the environmental contingencies has a lasting effect on behavior. The reactive model simply produces a change in performance, a behavioral modification that should disappear after the environmental contingencies are removed. In contrast, if the response-outcome relation is encoded, this learning should have a lasting effect on behavior that is evident when subjects are subsequently tested under common conditions. A third operation was suggested by the primary critic of the master-yoke paradigm, R. Church (1964). He argued that to determine whether a particular response-outcome relation underlies learning, the experimenter (rather than the master subject) needs to control the response-outcome relation. If the relation is critical, then disrupting response-outcome contiguity should eliminate learning. Our next three experiments provide evidence on each of these points (for additional evidence see Buerger, Eisenstein, and Reep, 1981).

In conducting these experiments, we also sought to address a number of other problems that have haunted past studies. The difficulty is that a variety of training variables (session duration, initial flexion force, flexion criterion) were not standardized. We addressed this problem by using constant session durations (30 min), and by equating both flexion force (by means of a strain gauge) and response criterion (contact electrode depth) prior to training. Past studies have also been criticized for only presenting the data from representative subjects and the use of questionable statistical techniques. We addressed these issues by using full factorial designs, reporting group means, and employing rigorous statistical methods (analysis of variance followed by appropriate post hoc comparisons).

Evidence for Instrumental Conditioning in Spinal Rats

In our first experiment, rats were tested 24 hrs after receiving a spinal transection at T2. After they were placed in the test apparatus (Figure 2.6), the shock electrodes were attached and shock intensity was adjusted to equate flexion force (0.6 N) across subjects. They then received either contingent shock (master), noncontingent shock (yoked), or nothing (unshocked) for 30 min. The impact of this training on flexion duration, time in solution, and flexion number is depicted on the left side of Figure 2.8. Master subjects exhibited progressively longer flexion durations as a function of training (top). As a consequence, they contacted the solution less (middle). Notice that yoked rats, that received exactly the same amount of shock, did not exhibit an increase in flexion duration and contacted the solution nearly as much as the unshocked controls. Notice too that master-yoke differences emerged even though both groups exhibited roughly the same number of responses (bottom) for the first 5 min of training.

At the end of training, flexion force was reassessed. For both master and yoked rats, more intense shocks were needed to elicit the same flexion force (change from baseline $\pm$ standard error of the mean: Master = 0.26$\pm$0.06 mA; Yoked = 0.30$\pm$0.09 mA; Unshocked = 0.04$\pm$0.02 mA). Thus, shock per se reduced the vigor of the flexion response. But this shock-induced habituation cannot account for the master-yoke difference because both groups exhibited a comparable level of habituation.

The fact that only contingent shock produced a systematic increase in flexion duration provided one indication that spinal cord neurons can support a form of instrumental learning. The possibility remained, however, that this difference reflected a performance effect driven by the distinct environmental contingencies. To address this possibility, we tested subjects under common conditions with response contingent shock applied to the same leg. If the performance differences were simply dictated by the current environmental contingency, any differences observed during training should disappear when subjects were exposed to the same response-outcome relation. A period of approximately 10 min separated training from testing, during which flexion force and contact electrode depth were re-equated across subjects. This step was taken to insure that any behavioral differences observed during testing were not attributable to differences in shock reactivity. All of the subjects then received 30 min of response contingent shock.

The results are depicted on the right side of Figure 2.8. Rats that were previously unshocked exhibited a gradual increase in flexion duration (top). Subjects that had previously experienced contingent shock (master) exhibited some savings, with flexion duration reaching asymptote within just a few minutes. Surprisingly, rats that had previously received noncontingent shock (yoked) did not exhibit an increase in flexion duration and, consequently, spent far more time contacting the solution (middle). Notice that the yoked group failed to learn despite the fact that they exhibited the highest rate of responding (bottom) and thus, repeatedly experienced the response-outcome relation. It appears that prior exposure to noncontingent shock induced a kind of learning deficit that made these subjects insensitive to the instrumental relation (cf Chopin and Bennett, 1975). Such a learning deficit is frequently observed in intact subjects, a phenomenon known as learned helplessness (Maier and Seligman, 1976).

Given a mechanical model, distinct environmental contingencies can yield performance differences. But from this perspective, we should be able to re-align performance by equating the contingencies. This did not occur. Instead, master rats exhibited some saving while yoked rats showed a profound learning deficit. Taken together, these results suggest that spinal systems are sensitive to response-outcome relations (Criterion 1), somehow encode this relation (Criterion 4), and that this learning has a lasting effect (Criterion 3).

Our next experiment sought further evidence that the response-outcome relation was encoded. This was accomplished by experimentally manipulating response-outcome contiguity as recommended by Church (1964). One group (0 msec delay) of spinalized rats received shock as soon as the contact electrode touched the solution and shock was terminated as soon as the electrode was lifted above the solution.

Other groups experienced a delay in <u>both</u> shock onset and offset. For one group, onset and offset were delayed by 50 msec. Another group experienced delays of 100 msec and the fourth group had onset and offset delayed by 200 msec. The results are illustrated in Figure 2.9A. As usual, rats that were trained without a delay (0 msec) exhibited progressively longer flexion durations as a function of training. Interposing a delay of 50 msec had a negligible effect on learning, but longer delays eliminated learning. This provides converging evidence that spinal cord neurons are sensitive to response-outcome relations.

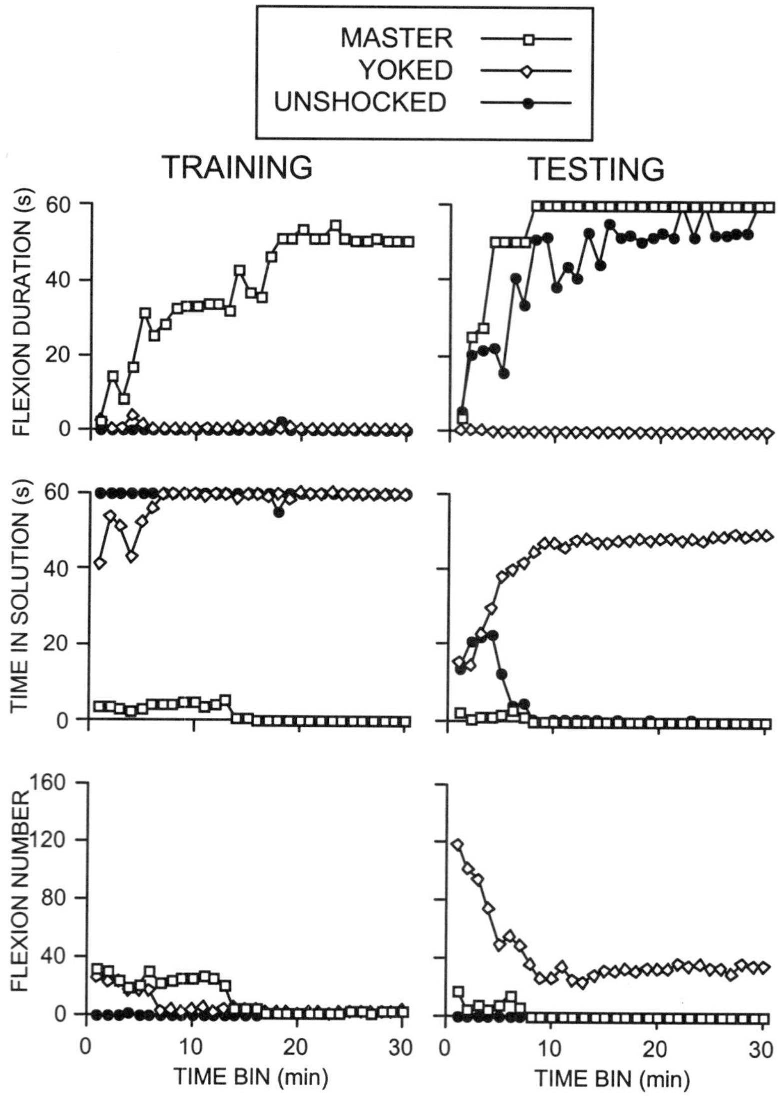

Figure 2.8. Flexion duration (top), time in solution (middle) and flexion number (bottom) over time for rats that were trained (left) with contingent shock (master), noncontingent shock (yoked), or remained unshocked. All subjects were then tested (right) by applying contingent to same leg. (Adapted from Grau et al., 1998, *Behavioral Neuroscience*, *112*, 1366-1386. © American Psychological Association.)

At the end of training, flexion force and contact electrode depth were re-equated and subjects were tested under common conditions with response contingent shock (0 msec delay). We again observed a learned-helplessness-like effect; rats that had failed to learn because shock onset and offset were delayed by 100-200 msec also failed to learn when the delay was removed.

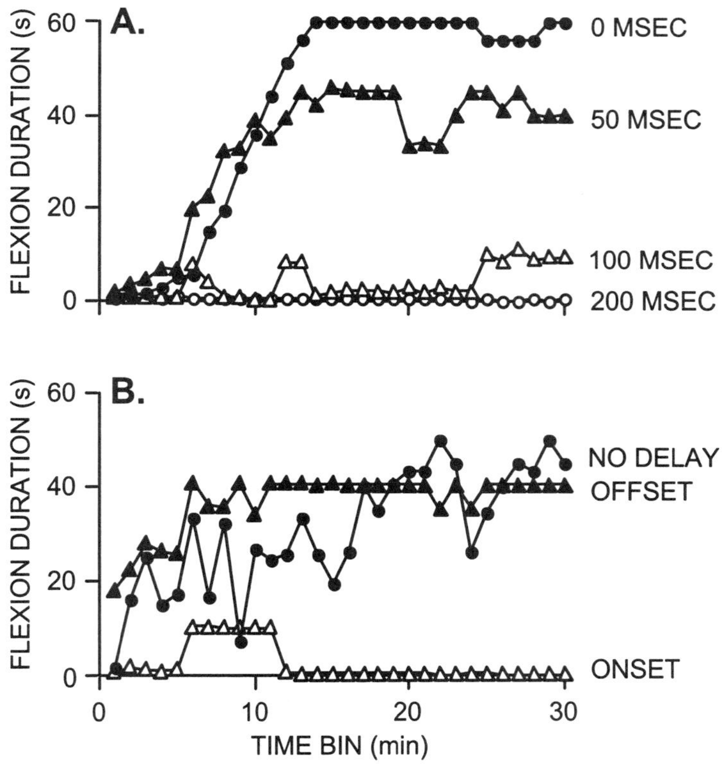

Figure 2.9. (A) Flexion duration over time in rats that received response contingent shock with both the onset and offset delayed by 0, 50, 100 or 200 msec. (B) Flexion duration over time in rats that received response contingent shock with no delay, onset delayed by 100 msec, or offset delayed by 100 msec. (Adapted from Grau et al., 1998, *Behavioral Neuroscience*, *112*, 1366-1386. © American Psychological Association.)

From the preceding experiment, it is not clear whether the critical outcome is shock onset or offset. Casually speaking, a leg extension could be punished by shock onset. Alternatively, contingent shock might produce an increase in flexion duration because an escape response is reinforced by shock offset. To evaluate these alternatives, a separate experiment was performed in which we independently manipulated shock onset and offset. One group had onset alone delayed by 100 msec while another had offset alone delayed by the same amount. A control group was trained in the usual fashion without a delay. We found that delaying shock offset had no effect (Figure 2.9B). In contrast, delaying onset eliminated learning. Abramson and Feinman (1987) obtained similar results in their studies of instrumental learning of eye extension in the green crab. At a mechanistic level, these results suggest that the simplest forms of instrumental learning may depend on the onset of a biologically significant stimulus for reinforcement. To derive reinforcements from the offset of such stimuli could require more sophisticated neural mechanisms. In vertebrates, this suggests that those forms of learning that depend on outcome offset (e.g., escape, signaled avoidance) require supraspinal mediation.

Evidence Learning Depends on Spinal Cord Neurons

So far, we have not presented any evidence that the learning in this paradigm is neurally mediated (Criterion 2). To address this issue, we microinjected lidocaine (50 μl, 20 μg/μl) into the lumbar region of the spinal cord. This local anesthetic disrupts the generation and conduction of nerve impulses through its effect on voltage sensitive Na^+ channels. Lidocaine eliminated learning (Figure 2.10) which suggests that the learning is neurally mediated.

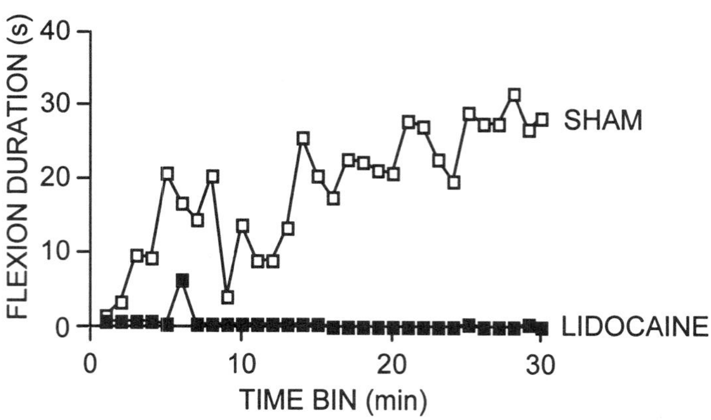

Figure 2.10. Flexion duration over time in spinal rats that were pretreated with lidocaine or nothing (SHAM) prior to training with contingent shock.

Evidence that the learning depends on spinal cord neurons was obtained from studies examining whether our behavioral effects would transfer to the contralateral leg (Joynes, Crown, Ferguson, and Grau, submitted). In the first experiment, rats received contingent shock (master) or noncontingent shock (yoked) to one hindleg. A control group remained unshocked during this training period. At the end of this training, we measured the shock intensity needed to elicit a flexion response of 0.6 N. As usual, shocked rats (both master and yoked) required more intense shock, but this effect was limited to the ipsilateral leg. Thus, it appears that the shock-induced decrease in flexion force reflected a response-specific effect, perhaps muscle fatigue. After shock intensity was adjusted to obtain equivalent flexion forces, both across legs and subjects, the response criterion (contact electrode depth) was equated. We then assessed learning by applying contingent shock to either the same (ipsilateral) or opposite (contralateral) leg. The results are illustrated in Figure 2.11. It is clear that nearly identical results were obtained irrespective of whether subjects were tested on the ipsilateral (left) or contralateral (right) leg. In both cases, the master and unshocked groups rapidly learned, exhibiting a progressive increase in flexion duration, and rats that had previously experienced noncontingent shock failed to learn. These results indicate that the deficit observed after noncontingent shock does not reflect a peripheral performance effect, such as muscle fatigue.

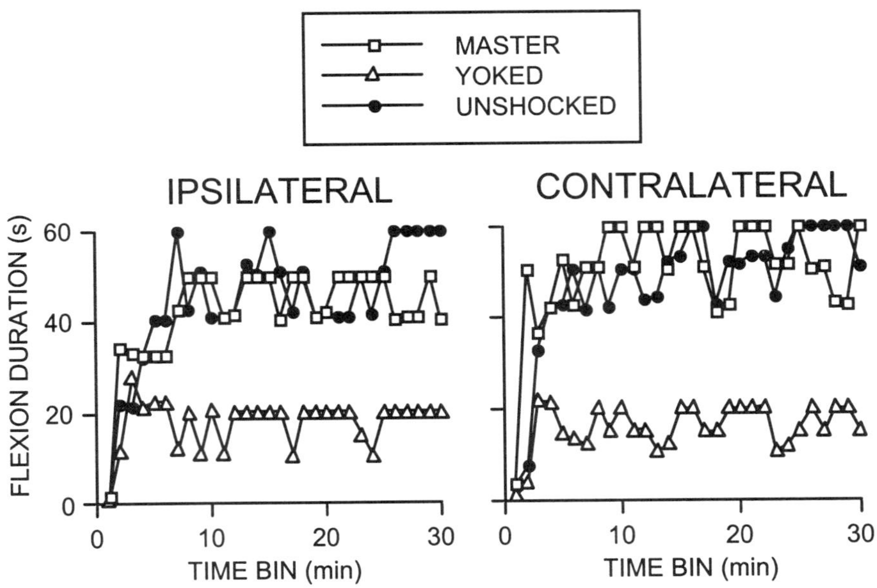

Figure 2.11. Flexion duration over time in spinal rats that had previously experienced contingent shock (master), noncontingent shock (yoked), or nothing (unshocked) to either the same (left) or opposite (right) leg.

It is not clear from the preceding experiment whether the beneficial effects of contingent shock transfer to the contralateral leg. Close inspection of the data revealed that master rats seem to learn more rapidly during the first 2 min of testing. This effect was equally robust on both test legs and was statistically significant. Yet, its transient nature raises concerns. It was possible that a transient effect was observed simply because the control subjects learned very rapidly, effectively masking the potential savings effect. To address this problem, Eric Crown designed an experiment in which he equated flexion force and response criterion prior to training. Half of the subjects received contingent shock while the other half served as the unshocked controls. He then tested subjects under common, but more difficult, conditions. This was achieved by adding an equal amount of solution to the container under each leg, an amount that would have submerged the contact electrode by an additional 4 mm at the start of training. Subjects received 30 min of testing with contingent shock applied to either the ipsilateral or contralateral leg. Notice that unshocked rats are being tested using a higher response criterion (greater contact electrode depth) than that employed in our preceding experiments. This should impede learning in the control subjects making it easier to resolve the potential beneficial effects of contingent shock. As shown in Figure 2.12, raising the response criterion did indeed disrupt learning in the unshocked controls. Rats that had previously experienced contingent shock were, however, able to learn and this was true irrespective of whether they were tested on the ipsilateral or contralateral leg. This finding extends our earlier observation of a training-induced savings effect and shows that this effect transfers to the contralateral leg.

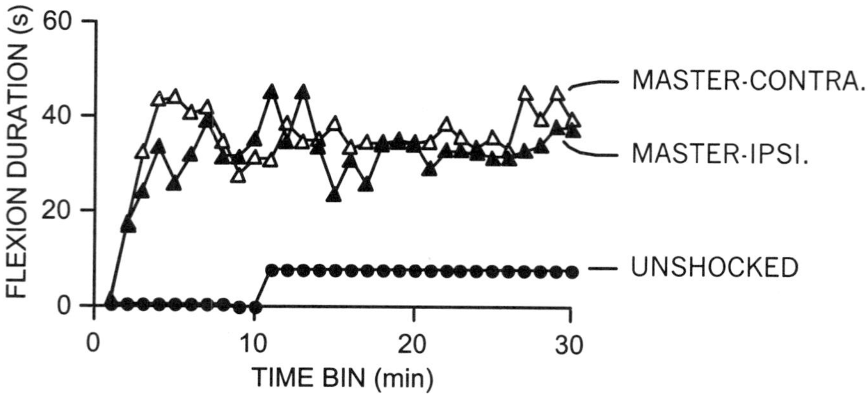

Figure 2.12. Flexion duration over time in spinal rats that had previously experienced contingent shock to the same (MASTER-IPSI) or opposite (MASTER-CONTRA) leg or nothing (UNSHOCKED). Prior to testing, the response criterion was made more difficult by adding 100 ml of solution.

Subsequent studies have attempted to identify the peripheral nerves that mediate learning. Because the sciatic nerve innervates the region in which shock is applied, Eric Crown and Adam Ferguson tested the impact of cutting this nerve near the dorsal root entry zone. They found that this physiological manipulation eliminated learning. This effect does not appear to be limited to a disruption of motor output, for cutting the sciatic nerve also prevented the development of the learning deficit. Specifically, rats that received noncontingent shock to a leg in which the sciatic nerve was cut did not exhibit a learning deficit when contingent shock was applied to the contralateral leg. Apparently, cutting the sciatic nerve prevented noncontingent shock from inducing the neurochemical/neurophysiological changes that underlie the learning deficit.

Neurochemical Mechanisms: Preliminary Observations

Learning Depends on the NMDA Receptor

We have begun to explore the neurochemical mechanisms that mediate learning in our instrumental paradigm. Our first experiments focused on the NMDA receptor. As discussed by Coderre, Willis, and Woolf, exposure to a noxious stimulus can sensitize neurons within the spinal cord, a phenomenon we will refer to as central sensitization (Coderre, Katz, Vaccarino, and Melzack, 1993; Wall and Woolf, 1984; Willis, Sluka, Rees, and Westlund, 1996; Woolf, 1983, 1984). Evidence suggests that this sensitization effect depends on the N-methyl-D-aspartate (NMDA) receptor, for it is blocked by NMDA antagonists (Dickenson and Sullivan, 1987; Ma and Woolf, 1995; Thompson, King, and Woolf, 1990; Willis et al., 1996). To evaluate whether this receptor is involved in our instrumental learning paradigm, we assessed the impact of the NMDA antagonist D,L-2,-2-amino-5-phosphonovalerate (APV) (Joynes, Janjua, and Grau, submitted). Separate groups received intrathecal injections (5 μl) of APV at a concentration of 0, 10, 20, or 40 mM. At these concentration levels, Cahusac, Evans, Hill, Rodriguez, and Smith (1984) showed that approximately 5.9% of the drug enters the spinal cord, yielding a concentration level of 12-60 μM. Comparable concentrations (40-50 μM) block hippocampal long-term potentiation (LTP) and spinal sensitization in vitro (Collingridge and Bliss, 1987; Thompson et al., 1990).

Subjects were given APV and placed in the test apparatus. The drug had no effect on the vigor of the flexion response and doses below 40 mM had no effect on initial flexion duration. The impact of the drug on learning is illustrated in Figure 2.13. Vehicle treated subjects exhibited the usual increase in flexion duration. The lowest dose of APV (10 mM) had no effect, but the two highest does (20-40 mM) eliminated learning.

While these results suggest an intriguing link to central sensitization and long-term potentiation, many important questions remain unanswered. One concerns an unusual characteristic of spinal sensitization: NMDA antagonists can disrupt its maintenance as well as induction (Woolf and Thompson, 1991). In a similar fashion, we found intrathecal APV (20 mM) administered after 30 min of training with contingent shock disrupted performance of the instrumental response. This pattern of results contrasts with those obtained in hippocampal preparations wherein NMDA antagonists disrupt the induction, but not the maintenance of LTP (e.g., Staubli, Thibault, DiLorenzo, and Lynch, 1989). A second quandary stems from our stimulus parameters: central sensitization appears to require C-fiber activation (Davis and Lodge, 1987; Dickenson and Sullivan, 1987), yet it is not clear whether our relatively mild stimuli engage C-fibers. Perhaps APV is effective in our paradigm because it affects higher order neurons of polysynaptic reflex pathways responsive to A-fiber stimulation (Davies and Watkins, 1983).

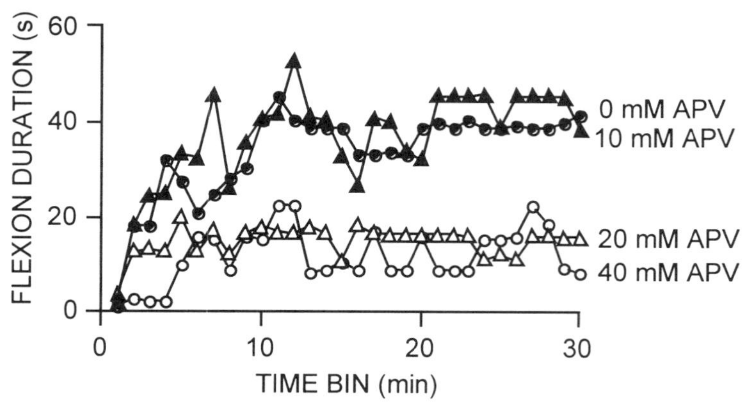

Figure 2.13. Flexion duration over time in spinal rats that received anintrathecal injection of 0, 10, 20, or 40 mM APV prior to training with contingent shock.

An Opioid-Mediated Process Prevents Learning

Other studies have begun to examine the neurochemical systems involved in the learning deficit. In the preceding experiments, a deficit was induced by yoking rats to master subjects. Although this approach allowed us to address some learning-related questions, it introduced a source of uncontrolled variability. Because each master subject learned at a different rate, each yoked subject received a different amount and distribution of shock. To eliminate this source of variability, we developed a paradigm in which we, rather than the master subject, controlled the application of noncontingent shock.

Our first experiments were parametric in nature and were designed to elucidate the shock parameters needed to induce a learning deficit. Shock duration was set to a value similar to that produced by a typical master subject (80 msec) and shocks were presented on a variable interval that ranged from 0.2 to 3.8 s (mean interstimulus interval = 2 s). Rats that experienced this shock schedule for just 6 min exhibited a robust learning deficit on the contralateral leg and this effect lasted 20 hrs (Crown, Joynes, Ferguson, and Grau, submitted).

Using this procedure, we examined whether the learning deficit was, in part, opioid mediated (Joynes and Grau, submitted). These studies were motivated by experiments in our laboratory demonstrating that exposure to noncontingent tailshock can induce an antinociception on the tail-flick test in spinal rats that is blocked by the opioid antagonist naltrexone (Meagher et al., 1993; Grau, Burks, Kallina, King, and Meagher, 1996). We were also intrigued by studies linking opioid peptides to the behavioral deficits observed after noncontingent shock in intact rats (Drugan and Maier, 1983; Whitehouse, Walker, Margules, and Bersh, 1983). In our first study, spinal rats were given an intrathecal injection (1 μl) of naltrexone at a dose of 0, 3.5, 7.0, or 14.0 μg/μl. Half of the subjects in each drug condition then received 6 min of noncontingent legshock. The remaining subjects served as the unshocked controls. Rats were then tested by applying contingent shock to the contralateral leg. The results are illustrated in Figure 2.14A. As usual, exposure to noncontingent shock disrupted learning in the vehicle control. More importantly, this deficit was eliminated by 7-14 μg/μl of naltrexone.

Naltrexone pretreatment could have eliminated the learning deficit by either: 1) preventing its induction during noncontingent shock exposure; or 2) preventing the expression of an opioid-mediated deficit at the time of testing. To evaluate these alternatives, a second experiment was conducted in which subjects received 7.0 μg/μl of naltrexone or its vehicle. Half of the subjects in each drug condition then received 6 min of noncontingent shock while the remaining subjects served as the unshocked controls. To allow the drug to clear the system, subjects were tested 20 hrs later. Prior to testing, half of the subjects in each condition received 7.0 μg/μl of naltrexone and the other half received the vehicle. All subjects then received 30 min of contingent shock applied to the contralateral leg. The results are illustrated in Figure 2.14B. Noncontingently shocked rats that received saline prior to training and testing (Group S-S) exhibited a learning deficit. Administering naltrexone prior to training alone (Group N-S) did not prevent the induction of the learning deficit. The deficit was, however, eliminated when naltrexone was given prior to testing (Groups S-N and N-N). It appears that the opioid antagonist was effective because it prevented the expression of the learning deficit rather than its induction.

Though naltrexone has a higher affinity for the mu receptor, it also acts on delta and kappa receptors (Millan, 1986). Given this, we examined the impact of the opioid antagonists Cys-Try-Orn-Pen-amide (CTOP), naltrindole hydrochloride (n-DOL), and nor-binaltorphimine dihydrochloride (n-BNI) which act selectively at the mu, delta, and kappa receptor, respectively. The drugs were dissolved in DMSO and delivered intrathecally (1 μl), at a dose that was equimolar to 7.0 μg/μl of naltrexone (18.5 nM), immediately after subjects received 6 min of noncontingent shock or nothing. Subjects were then tested by applying contingent shock to the contralateral leg. As usual, spinal rats that were previously unshocked exhibited longer flexion durations during testing (see Figure 2.14C). Vehicle treated subjects that had received noncontingent shock failed to learn. Only n-BNI eliminated this learning deficit, which suggests that a ligand that acts on the kappa opioid receptor was involved (cf Faden, 1990). Future studies will examine the relative dose response profiles of these effects and evaluate whether administration of a kappa agonist is sufficient to induce a learning/performance deficit.

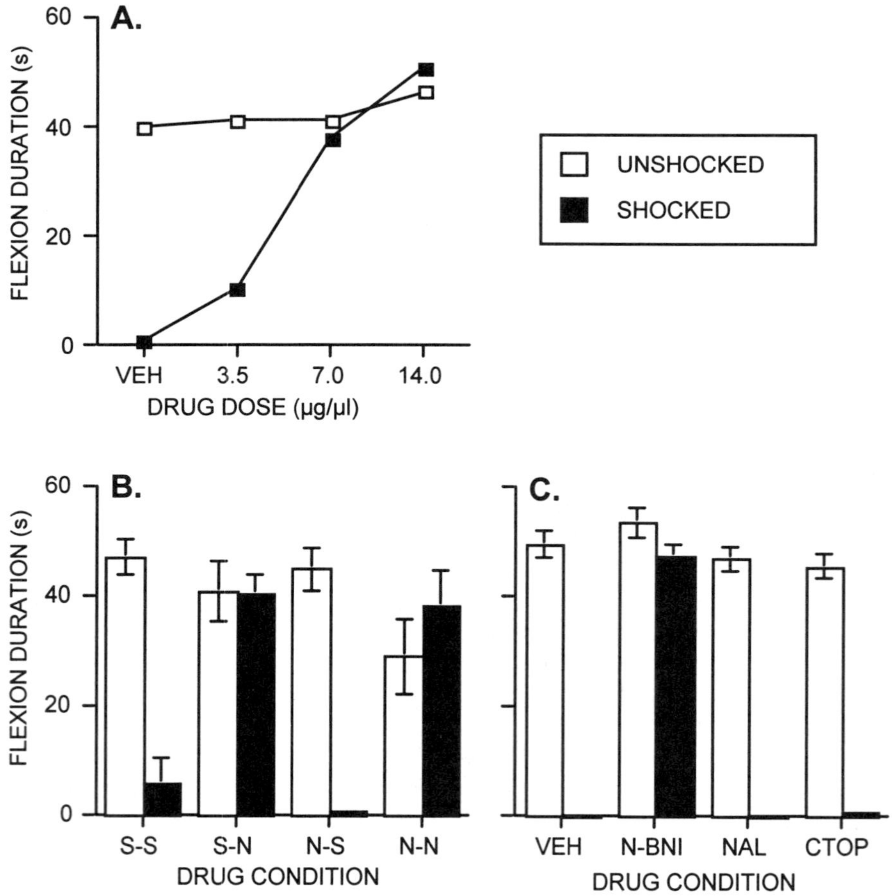

Figure 2.14. Mean flexion duration, collapsed across the 30 min of testing, in spinal rats tested with contingent shock. Subjects had previously received 6 min of noncontingent shock (SHOCKED) or nothing (UNSHOCKED). (A) Rats were pretreated with intrathecal naltrexone at a dose of 0 (VEH), 3.5, 7.0 or 14.0 µg/µl naltrexone prior to receiving noncontingent shock (filled symbols) or nothing (open symbols). Flexion force was then reset and subjects were tested with contingent shock applied to the contralateral leg. (B) Rats received saline (S) or naltrexone (N) prior to 6 min of noncontingent shock (filled bars) or nothing (open bars). Twenty hours later, half the subjects in each drug condition received naltrexone and the remaining subjects were given saline. They were then tested by applying contingent shock to the contralateral leg. (C) Rats received 6 min of noncontingent shock (filled bars) or nothing (open bars) followed by an intrathecal injection of either the vehicle (VEH), nor-BNI (N-BNI), naltrindole (NAL) or CTOP. Rats were then tested by applying contingent shock to the contralateral leg. In both (B) and (C) the error bars indicate the standard error of the mean.

Implications and Limits

Our second set of experiments examined the mechanisms that underlie the modification of a flexion response in spinal rats. At issue was whether this learning could be accounted for in terms of a reactive model, a mechanical system that might involve a form of single stimulus learning (e.g., sensitization) but is insensitive to response-outcome relations. The reactive model accurately anticipated that master rats would spend less time with their leg extended contacting the salt solution. However, it did not predict other important details: a) that master rats exhibit longer flexion durations; b) that flexion duration increases as a function of training; c) that master-yoke differences would emerge under conditions that produce similar response rates (flexions/minute); d) that delaying shock onset eliminates learning; and e) that the consequences of the environmental contingencies extend beyond their operational period. Only the instrumental account anticipated properties a-e, and we take this as strong evidence that systems within the spinal cord are sensitive to response-outcome relations.

The paradigm we used allowed us to evaluate sensitivity to the response-outcome relation without knowing, a priori, what aspect of the outcome was critical and when it was effective. On the basis of our results, we now know that the critical outcome is shock onset and that the impact of this outcome depends on leg position (the critical response). The result of this contingency is an increase in flexion duration.

The behavior observed in this paradigm bears an interesting similarity to an example of learning described by de Leon and Edgerton (de Leon, Hodgson, Roy, and Edgerton, in press; Edgerton, Roy, de Leon, Tillakarantne, and Hodgson, 1997; Hodgson, Roy, de Leon, Dobkin, and Edgerton, 1994). In exploring the motor systems that mediate stepping in spinal cats, they discovered that the stepping response can be modified through experience. If a rod is positioned so that the dorsal surface of the cat's paw hits it during the swing phase of walking, the cat learns to lift its leg sooner, lessening the impact against the rod. Thus, the onset of a nociceptive stimulus (striking the rod) during stepping causes the cat to exhibit a stronger flexion response, just as the onset of response-contingent shock augments flexion duration in spinal rats.

At the start of this section, we outlined a set of criteria that would need to be met in order to demonstrate that a behavioral phenomenon represents an instance of instrumental learning. From our results, it is clear that: 1) instituting a response-outcome relation produces a change in behavior (Criterion 1); 2) that this learning depends on neurons within the spinal cord (Criterion 2); 3) exposure to either contingent or noncontingent shock had an effect that extends beyond the training period (Criterion 3); and 4) that learning depends on the temporal relationship between the response and outcome (Criterion 4). The last 3 criteria provide evidence that the response-outcome relation is somehow neurally encoded.

The term instrumental learning has its roots in the reflexive tradition of Thorndike and Hull (Hilgard and Marquis, 1940). From this perspective, instrumental learning reflects just another form of elicited behavior, one that depends on the relationship established between a response and an outcome (the reinforcer). A very different perspective was advanced by Skinner (1937, 1938) who noted that organisms often do not appear to respond in a stereotyped, reflexive, fashion but instead operate on their environment in a flexible manner. To Skinner, such behavior was emitted, not elicited, and such behavior could not be accounted for by the reflexive principles advanced by Thorndike and Hull. Skinner distinguished this form of learning by introducing a new term, *operant* learning. The flexible nature of operant behavior suggests that neither the nature of the behavioral change (Criterion 5) nor the reinforcer (Criterion 6) are highly constrained.

A good example of operant learning is provided by Wolpaw's studies of the spinal stretch reflex (Carp and Wolpaw, 1994, 1995; Hatler, Carp, and Wolpaw, 1995; Wolpaw, Braitman, and Seegal, 1983; Wolpaw and Carp, 1990; Wolpaw and Lee, 1989). As detailed in this volume, his work has shown that monkeys can learn to modify the stretch reflex (by increasing or decreasing it) for an appetitive reinforcer (e.g., squirt of juice). Like our instrumental paradigm, the results of learning (the memory) appear to be stored within the spinal cord. However, in Wolpaw's paradigm, the processes that mediate the learning (the derivation of the response-outcome relation) are supraspinally mediated and these processes appear far more flexible, allowing for a range of behavioral outcomes (Criterion 5) and reinforcers (Criterion 6).

We have purposely used the term instrumental, rather than operant, to describe our behavioral paradigm for two reasons. First, our work follows in the reflexive tradition of Thorndike and Hull (Hilgard and Marquis, 1940). Second, the behavior we observe seems to lack the flexibility that stands as a hallmark of operant learning. Those familiar with the learning literature know that the second point has been tempered in recent years by the many demonstrations of constraints on learning (Bolles, 1970; Timberlake and Lucas, 1989). Nonetheless, the learning inherent within the spinal cord seems more highly prepared and, as such, may generally fail the advanced criteria for operant learning (5 and 6). Regarding Criterion 6, the plasticity observed within the spinal cord may be limited to aversive (potentially noxious) stimuli. Such a perspective is consistent with the notion that the simplest forms of learning evolved from mechanisms designed to detect and respond to injury (Walters, 1994). As to Criterion 5, others have attempted to address this issue by demonstrating spinal rats can learn to extend their leg when shock is applied at a different location (Sherman, Hoehler, and Buerger, 1982). However, there is no evidence that spinal rats can learn the opposite behavior (to maintain a flexion) when shock is applied at this new location. To truly address Criterion 5, a preparation is needed in which either an increase or decrease in the response can be arbitrarily shaped depending on when, not where, the outcome (shock onset) is applied. There is currently no evidence that spinal systems have this capacity.

Having demonstrated that spinal cord systems are capable of instrumental conditioning, further studies are needed to characterize the underlying mechanisms. We have suggested that the nominal response is leg position, a property that spinal cord systems could monitor through proprioceptive feedback. Through learning, components of this afferent signal could acquire the capacity to maintain the flexion response. Such a mechanism was suggested by Konorski and Miller (1937 a and b), the first to recognize the Pavlovian/instrumental distinction (what they referred to as Type I and Type II learning). Interestingly, reinforcement could select the afferent components (position) by means of pairing specific enhanced sensitization (wherein shock onset selectively strengthens a pre-existing reflex just as the facilitory neuron appears to strengthen the siphon-gill withdrawal reflex in *Aplysia* [Kandel and Schwartz, 1982]). Thus, our studies of instrumental learning suggest that additional mechanisms are operating within the spinal cord, systems that cannot be accounted for in terms of protection from habituation.

2.5 CONCLUSIONS

A half century ago, the field of learning was dominated by generalists who attempted to build theories of behavior using a unitary mechanism (associative learning) that could be flexibly applied to form new units of behavior (e.g., stimulus-response associations). This perspective was challenged in the 1960s by Garcia, the Brelands, and others who recognized that what the organism can learn is often highly constrained (Bolles, 1970; Breland and Breland, 1961; Garcia and Koelling, 1966; Garcia, Brett, and Rusiniak, 1989; Rozin and Kalat, 1971; Seligman, 1970; Timberlake and Lucas, 1989). Garcia even took the heretical step of suggesting that organisms have evolved specialized learning mechanisms designed to solve particular environmental problems (e.g., to span long intervals between taste and illness). Over the next decade, the generalists fought back, providing evidence that the same principles appear to govern learning in a variety of paradigms (Domjan, 1983). At the same time, the list of exceptions has grown (Garcia et al., 1989; Timberlake and Lucas, 1989).

Our work suggests an intermediate perspective that acknowledges from the start that a single mechanism cannot account for all of learning. For example, temporal relations between stimuli can be encoded by protection from habituation, pairing specific enhanced sensitization, or associative learning. Each set of mechanisms may share some basic features, as suggested by the generalists (e.g., the principles of extinction and cue competition seem to generalize across all systems capable of encoding Pavlovian relations). At the same time, the claim that these mechanisms can be operationally distinguished suggests that each system has its own set of unique features. Thus, we expect learning phenomena that depend on distinct mechanisms to exhibit qualitative differences, as Garcia suggested. The difference is that we assume that these mechanisms evolved, not to solve a particular species-specific problem, but instead, to solve a class of problems. From this perspective, the same functional mechanism, perhaps the same biological mechanism, may serve many masters. Moreover, even within a single preparation, multiple mechanisms may be brought to bear, with the relative contribution of each varying as a function of training, development, and neural complexity (Joynes and Grau, 1996; Sahley and Ready, 1988).

Our aim has been to characterize the functional mechanisms that underlie learning within the spinal cord. Our work confirms that these systems are sensitive to the environmental relations that underlie Pavlovian and instrumental conditioning. At the same time, we have noted that these processes appear to lack some of the flexibility and complexity that characterizes the most advanced forms of infrahuman learning. Spinal cord systems are necessarily highly prepared whereas supraspinally mediated behavior may be either prepared or unprepared. The latter allows the brain to flexibly map new environmental relations across time irrespective of the stimulus modality, response form, or affective state. Lacking this flexibility, spinal cord systems may be unable to solve a variety of environmental problems (e.g., negative patterning, trace conditioning, Sidman avoidance). However, this remains an empirical issue and we would be the first to admit that we have underestimated the capacity of spinal cord systems in the past.

ACKNOWLEDGEMENTS

We would like to thank Eric Crown, Michael Domjan, Adam Ferguson, Mary Meagher, and Amy Sieve for their comments on an earlier version of this paper. This work was funded by National Institute of Mental Health Grants MH48994 and MH60157. Correspondence concerning this article should be addressed to James W. Grau, Department of Psychology, Texas A and M University, College Station, Texas, 77843. Electronic mail may be sent to j-grau@tamu.edu.

2.6 REFERENCES

Abramson, C.I., and Feinman, R.D. Operant punishment of eye elevation in the green crab, *Carcinus maenas*. *Behavioral and Neural Biology*, 1987; 48:259-277.

Alkon, D.L. Persistent calcium-mediated changes of identified membrane currents as a cause of associative learning. In D.L. Alkon and J. Farley (Eds.), *Primary Neural Substrates of Learning and Behavioral Change*. (pp. 291-324). New York: Cambridge University Press, 1984.

Beggs, A.L., Steinmetz, J.E., and Patterson, M.M. Classical conditioning of a flexor nerve response in spinal cats: Effects of tibial nerve CS and a differential conditioning paradigm. *Behavioral Neuroscience*, 1985; 99:496-508.

Bolles, R.C. Species-specific defense reactions and avoidance learning. *Psychological Review*, 1970; 77:32-48.

Bower, G.H., and Hilgard, E.R. *Theories of learning*. Englewood Cliffs, N. J.: Prentice-Hall. 1981.

Breland, K., and Breland, M. The misbehavior of organisms. *American Psychologist*, 1961; 16:681-684.

Buerger, A.A., and Chopin, S. F. Instrumental avoidance conditioning in spinal vertebrates. *Advances in Psychobiology*, 1976; 3:437-461.

Buerger, A.A., Eisenstein, E.M., and Reep, R.L. The yoked control in instrumental avoidance conditioning: An empirical and methodological analysis. *Physiological Psychology*, 1981; 9:351-353.

Buerger, A.A., and Fennessy, A. Long-term alteration of leg position due to shock avoidance by spinal rats. *Nature*, 1970; 225:751-752.

Cahusac, P.M.B., Evans, R.H., Hill, R.G., Rodriguez, R.E., and Smith, D.A.S. The behavioral effects of an N-methylaspartate receptor antagonist following application to the lumbar spinal cord of conscious rats. *Neuropharmacology*, 1984; 23: 719-724.

Carp, J.S., and Wolpaw, J.R. Motorneuron plasticity underlying operantly conditioned decrease in primate H-reflex. *Journal of Neurophysiology*, 1994; 72:431-442.

Carp, J.S., and Wolpaw, J.R. Motorneuron properties after operantly conditioned increase in primate H-reflex. *Journal of Neurophysiology*, 1995; 73:1365-1373.

Chopin, S.F., and Bennett, M.H. The effect of unavoidable shock on instrumental avoidance conditioning in spinal rats. *Physiology and Behavior*, 1975; 14:399-401.

Chopin, S.F., and Buerger, A.A. Instrumental avoidance conditioning in the spinal rat. *Brain Research Bulletin*, 1976; 1:177-183.

Church, R.M. Systematic effect of random error in the yoked control design. *Psychological Bulletin*, 1964; 62:122-131.

Church, R.M. The yoked control design. In T. Archer and L. Nilsson (Eds.), *Aversion, avoidance, and anxiety: Perspectives on aversively motivated behavior*. Hillsdale, N. J.: Erlbaum, 1989.

Church, R.M., and Lerner, N.D. Does the headless roach learn to avoid? *Physiological Psychology*, 1976; 4:439-442.

Coderre, T.J. The role of excitatory amino acid receptors and intracellular messengers in persistent nociception after tissue injury in rats. *Molecular Neurobiology*, 1994; 7:229-246.

Coderre, T.J., Katz, J., Vaccarino, A.L., and Melzack, R. Contribution of central neuroplasticity to pathological pain: review of clinical and experimental evidence. *Pain*, 1993; 52:259-285.

Cohen, J. *Statistical power analysis for the behavioral sciences* (2nd ed.). Hillsdale, N. J.: Lawrence Erlbaum Associates, 1988.

Collingridge, G.L., and Bliss, T.V.P. NMDA receptors–their role in long-term potentiation. *Trends in Neuroscience,* 1987; 10:288-293.

Crown, E.D., Joynes, R.L., Ferguson, A.R. and Grau, J.W. (submitted). Instrumental learning within the spinal cord: IV. Induction and retention of the learning deficit observed after noncontingent shock.

Davies, S. N., and Lodge, D. Evidence for involvement of N-methyl-aspartate receptors in 'wind-up' of class 2 neurons in the dorsal horn of the rat. *Brain Research*, 1987; 424:402-406.

Davies, J., and Watkins, J.C. Role of excitatory amino acid receptors in mono- and polysynaptic excitation in the cat spinal cord. *Experimental Brain Research*, 1983; 49:280-290.

Davis, M. Effects of ISI length and variability on startle response habituation to different intensities of tone. *Psychonomic Science*, 1970; 12:337-338.

de Leon, R.D., Hodgson, J.A., Roy, R.R., and Edgerton, V.R. (in press). Evidence for learning in the adult mammalian spinal cord: Use-dependent recovery of locomotion in spinal cats. *Journal of Neuroscience.*

Dickenson, A.H., and Sullivan, A.F. Evidence for a role of the NMDA receptor in the frequency dependent potentiation of deep rat dorsal horn nociceptive neurons following C-fibre stimulation. *Neuropharmacology*, 1987; 26:1235-1238.

Domjan, M. Biological constraints on instrumental and classical conditioning: Implications for general process theory. In G.H. Bower (Ed.), *The psychology of learning and motivation* (Vol. 17). New York: Academic Press, 1983.

Domjan, M. *The principles of learning and behavior*. Pacific Grove, California: Brooks/Cole Publishing, 1993.

Domjan, M. Learning: Overview. In A.E. Kazdin (Ed.), *Encyclopedia of Psychology*. Washington, D.C.: American Psychological Association, and New York: Oxford University Press, 2000

Drugan, R.C., and Maier, S.F. Analgesic and opioid involvement in the shock-elicited activity and escape deficits produced by inescapable shock. *Learning and Motivation*, 1983; 14:30-47.

Durkovic, R.G. Classical conditioning, sensitization, and habituation in the spinal cat. *Physiology and Behavior*, 1975; 14: 297-304.

Edgerton, V.R., Roy, R.R., de Leon, R., Tillakaratne, and Hodgson, J. A. Does motor learning occur in the spinal cord? *Neuroscientist,* 1997; 3:287-294.

Eisenstein, E.M. A comparison of activity and position response measures of avoidance learning in the cockroach, *P. americana. Brain Research*, 1970; 21:143-147.

Faden, A.I. Opioid and nonopioid mechanisms may contribute to dynorphin's pathophysiological actions in spinal cord injury. *Annals of Neurology*, 1990; 27:67-74.

Fanselow, M.S. Conditioned fear-induced opiate analgesia: A competing motivational state theory of stress analgesia. *Annals of the New York Academy of Sciences*, 1986; 467:40-54.

Fitzgerald, L.A., and Thompson, R. F. Classical conditioning of the hindlimb flexion reflex in the acute spinal cat. *Psychonomic Science*, 1967; 47:345-351.

Garcia, J., Brett, L.P., Rusiniak, K.W. Limits of Darwinian conditioning. In S.B. Klein and R.R. Mowrer (Eds.), *Contemporary learning theories: Instrumental conditioning theory and the impact of biological constraints on learning* (pp. 237-275). Hillsdale, N. J.: Erlbaum, 1989.

Garcia, J., and Koelling, R. A. Relation of cue to consequence in avoidance learning. *Psychonomic Science*, 1966; 4:123-124.

Gibbon, J., and Balsam, P.D. Spreading association in time. In C.M. Locurto, H.S. Terrace, and J. Gibbon (Eds.), *Autoshaping and conditioning theory* (pp. 219-253). New York: Academic Press, 1981.

Gormezano, I., and Kehoe, E.J. Classical conditioning: Some methodological-conceptual issues. In W.K. Estes (Ed.), *Handbook of learning and cognitive processes*, Vol. 2 (pp. 143-179). Hillsdale, N. J.: Erlbaum, 1975.

Grau, J.W. The central representation of an aversive event maintains the opioid and nonopioid forms of analgesia. *Behavioral Neuroscience,* 1987; 101:272-288.

Grau, J.W., Barstow, D.G., and Joynes, R.L. Instrumental learning within the spinal cord. I. behavioral properties. *Behavioral Neuroscience,* 1998; 112:1366-1386.

Grau, J.W., Burks, K., Kallina, C.F., King, T.E., and Meagher, M.W. Activation of the opioid and nonopioid antinociceptive systems in pentobarbital anesthetized rats: Assessing the role of shock severity. *Psychobiology*, 1996; 24:71-84.

Grau, J.W., Salinas, J.A., Illich, P.A., and Meagher, M.W. Associative learning and memory for an antinociceptive response in the spinalized rat. *Behavioral Neuroscience*, 1990; 104:489-494.

Groves, P.M., Lee, D., and Thompson, R.F. Effects of stimulus frequencey and intensity on habituation and sensitization in acute spinal cat. *Physiology and Behavior*, 1969; 4:383-388.

Halter, J.A., Carp, J.S., and Wolpaw, J.R. Operantly conditioned motorneuron plasticity: Possible role of sodium channels. *Journal of Neurophysiology*, 1995; 73:867-871.

Hawkins, R.D., Abrams, T.W., Carew, T.J., and Kandel, E.R. A cellular mechanism of classical conditioning in Aplysia: Activity dependent amplification of presynaptic facilitation. *Science*, 1983; 219:400-405.

Hawkins, R.D., and Kandel, E.R. Is there a cell-biological alphabet for simple forms of learning? *Psychological Review,* 1984; 91:375-391.

Hearst, E. The classical-instrumental distinction: Reflexes, voluntary behavior, and categories of associative learning. In W. K. Estes (Ed.), *Handbook of learning and cognitive processes: Conditioning and behavior theory.* Hillsdale, N. J.: Erlbaum, 1975.

Hilgard, E.R., and Marquis, D.G. *Conditioning and learning.* New York: Appleton-Century-Crofts, 1940.

Hodgson, J.A., Roy, R.R., de Leon, R., Dobkin, B., and Edgerton, V.R. Can the mammalian lumbar spinal cord learn a motor task? *Medicine and Science in Sports and Exercise,* 1994; 26:1491-1947.

Holland, P.C. Event representation in Pavlovian conditioning: Image and action. *Cognition,* 1990; 37: 105-131.

Horridge, G.A. (1962). Learning of leg position by the ventral nerve cord in headless insects. *Proceedings of the Royal Society,* 1962; 157:33-52.

Hull, C.L. *Principles of behavior.* New York: Appleton-Century-Crofts, 1943.

Humphrey, G. *The nature of living in its relation to the living system* (pp. 165-179). New York: Harcourt, Brace and Co, 1933.

Illich, P.A., Salinas, J.A., and Grau, J.W. Latent inhibition, overshadowing, and blocking of a conditioned antinociception response in spinalized rats. *Behavioral and Neural Biology,*1994; 62: 140-150.

Joynes, R.L., Crown, E.D., Ferguson, A.R. and Grau, J.W. (submitted). Instrumental learning within the spinal cord: V. Evidence for central mediation.

Joynes, R.L. and Grau, J.W. (submitted). Instrumental learning with the spinal cord: III. Prior exposure to noncontingent shock induces a behavioral deficit that is blocked by an opiod antagonist.

Joynes, R.L., and Grau, J.W. Mechanisms of Pavlovian conditioning: The role of protection from habituation in spinal conditioning. *Behavioral Neuroscience*, 1996; 110:1375-1387.

Joynes, R.L., Illich, P.A., Grau, J.W. Evidence for spinal conditioning in intact rats. *Neurobiology of Learning and Memory*, 1997; 67:64-68.

Joynes, R.L., Janjua, K.R., and Grau, J.W. (submitted). Instrumental learning within the spinal cord: II. The NMDA antagonist AP5 disrupts the acquisition and maintenance of the instrumental response.

Kandel, E.R., and Schwartz, J.H. Molecular biology of learning: Modification of transmitter release. *Science*, 1982; 218:433-442.

Kimble, G.A. *Hilgard and Marquis' conditioning and learning.* New York: Appleton-Century-Crofts, 1961.

Konorski, J.A. *Integrative activitity of the brain*. Chicago: University of Chicago Press. 1967.

Konorski, J.A., and Miller, S.M. On two types of conditioned reflex. *Journal of General Psychology,* 1937;16:264-273.

Konorski, J.A., and Miller, S.M. Further remarks on two types of conditioned reflex. *Journal of General Psychology*, 1937; 16:405-407.

Lubow, R.E., Schnur, P., and Rifkin, B. Latent inhibition and conditioned attention theory. *Journal of Experimental Psychology: Animal Behavior Processes*, 1976; 2:163-174.

Ma, Q..P., and Woolf, C.J. Noxious stimuli induce an N-methyl-D-aspartate receptor-dependent hypersensitivity of the flexion withdrawal reflex to touch: implications for the treatment of mechanical allodynia. *Pain*, 1995; 61:383-390.

Mackintosh, N.J. *The psychology of animal learning*. New York: Academic Press, 1974.

Maier, S.F. Determinants of the nature of environmentally induced hypoalgesia. *Behavioral Neuroscience*, 1989; 103:131-143.

Maier, S.F. and Seligman, M.E.P. Learned helplessness: Theory and evidence. *Experimental Psychology*, 1976; 105:3-46.

Meagher, M.W., Chen, P.S., Salinas, J.A., and Grau, J.W. Activation of the opioid and nonopioid hypoalgesic systems at the level of the brainstem and spinal cord: Does a coulometric relation predict the emergence or form of environmentally induced hypoalgesia? *Behavioral Neuroscience*, 1993; 107:493-505.

Millan, M.J. Multiple opioid systems and pain. *Pain*, 1986; 27:303-348.

Miller, R.R., and Matzel, L.D. Contingency and relative associative strength. In S. B. Klein and R. R. Mowrer (Eds.), *Contemporary learning theories: Pavlovian conditioning and the status of traditional learning theory* (pp. 61-84). Hillsdale, N.J.: Erlbaum, 1989.

Mitchell, D., Scott, D.W., and Mitchell, L.K. Attenuated and enhanced neophobia in the taste-aversion "delay of reinforcement" effect. *Animal Learning and Behavior,* 1977; 5:99-102.

Patterson, M.M., Cegavske, C.F., and Thompson, R.F. Effects of a classical conditioning paradigm on hind-limb flexor nerve response in the immobilized cat. *Journal of Comparative and Physiological Psychology*, 1973; 84:88-97.

Pavlov, I.P. Conditioned reflexes [G. V. Anrep, trans.]. London: Oxford University Press, 1927.

Pfautz, P.L., Donegan, N.H., and Wagner, A.R. Sensory preconditioning versus protection from habituation. *Journal of Experimental Psychology: Animal Behavior Processes*, 1978; 4:286-295.

Prokasy, W.F., Grant, D.A., and Myers, N.A. Eyelid conditioning as a function of UCS intensity and intertrial interval. *Journal of Experimental Psychology*, 1958; 55:242-246.

Rescorla, R.A. Comments on three Pavlovian paradigms. In D.L. Alkon and J. Farley (Eds.), *Primary neural substrates of learning and behavioral change.* Cambridge: Cambridge University Press. 1984.

Rescorla, R.A. Behavioral studies of Pavlovian conditioning. *Annual Review of Neuroscience*, 1988; 11: 329-352.

Rescorla, R.A., and Wagner, A.R. A theory of Pavlovian conditioning: Variations in the effectiveness of reinforcement and nonreinforcement. In A. H. Black and W.F. Prokasy (Eds.), *Classical conditioning II. Current theory and research.* New York: Appleton-Century-Crofts, 1972.

Riley, A.L. In response to and in defense of Mitchell and Revusky: An analysis of nonassociative effects. *Animal Learning and Behavior,* 1978; 6:472-473.

Rozin, P., and Kalat, J.W. Specific hungers and poison avoidance as adaptive specializations of learning. *Psychological Review*, 1971; 78:459-486.

Sahley, C.L., and Ready, D.F. Associative learning modifies two behaviors in the leech, *Hirudo medicinalis. Journal of Neuroscience*, 1988; 8:4612-4720.

Satinoff, E. Neural organization of evolution of thermal regulation in mammals. *Science,* 1978; 201: 16-22.

Seligman, M.E.P. On the generality of the laws of learning. *Psychological Review,* 1970; 77:406-418.

Sherman, B.S., Hoehler, F.K., and Buerger, A.A. Instrumental avoidance conditioning of increased leg lowering in the spinal rat. *Physiology and Behavior*, 1982; 25:123-128.

Sherrington, C.S. *The integrative action of the nervous system.* New Haven: Yale University Press, 1906.

Shurrager, P.S., and Culler, E. Conditioning in the spinal dog. *Journal of Experimental Psychology*, 1940; 26: 133-159.

Skinner, B.F. Two types of conditioned reflex: A reply to Konorski and Miller. *Journal of General Psychology*, 1937; 16:272-279.

Skinner, B.F. *The behavior of organisms.* Englewood Cliffs, N. J.: Prentice-Hall, 1938.

Solomon, P.R., Vander Schaaf, E.R., Thompson, R.F., and Weisz, D.J. Hippocampus and trace conditioning in rabbit's classically conditioned nictitating membrane response. *Behavioral Neuroscience,* 1986; 100:729-744.

Spence, K.W. and Norris, E.B. Eyelid conditioning as a function of inter-trial interval. *Journal of Experimental Psychology,* 1950; 40:716-720.

Squire, L.R., and Zola, S.M. Structure and function of declarative and nondeclarative memory systems. *Proceedings of the National Academy of Sciences*, 1996; 93:13515-13522.

Staubli, U., Thibault, O., DiLorenzo, M., and Lynch, G. Antagonism of NMDA receptors impairs acquisition but not retention of olfactory memory. *Behavioral Neuroscience*, 1989; 103:54-60.

Sutherland, R.J., and Rudy, J.W. Configural association theory: The role of the hippocampal formation in learning, memory, and amnesia. *Psychobiology*, 1989; 17:129-144.

Thompson, S.W.N., King, A.E., and Woolf, C.J. Small caliber afferents produce a heterosynaptic facilitation of the synaptic response evoked by primary A fibres in the neonatal rat spinal cord in vitro. *Journal of Neurophysiology,* 1990; 69:2116-2128.

Thorndike, E.L. Animal intelligence: An experimental study of associative processes in animals. *Psychological Review, Monograph*, 1898; 2 (8).

Timberlake, W., and Lucas, G.A. Behavior systems and learning: From misbehavior to general principles. In S.B. Klein and R. R. Mowrer (Eds.), *Contemporary learning theories: Instrumental conditioning theory and the impact of biological constraints on learning* (pp. 237-275). Hillsdale, N. J.: Erlbaum, 1989.

Wagner, A.R. Priming in STM: An information-processing mechanism for self-generated or retrieval-generated depression in performance. In T.J. Tighe and R.N. Leaton (Eds.), *Habituation: Perspectives from child development, animal behavior, and neurophysiology* (pp. 95-128). Hillsdale, N.J.: Erlbaum, 1976.

Wall, P.D., and Woolf, C.J. Muscle but not cutaneous C-afferent input produces prolonged increases in the excitability of the flexion response in the rat. *Journal of Physiology*, 1984; 356:443-458.

Walters, E.T., Injury related behavior and neuronal plasticity: An evolutionary perspective on sensitization, hyperalgesia, and analgesia. *International Review of Neurobiology,* 1994; 36:325-426.

Walters, E.T., Byrne, J.H. Associative conditioning of single sensory neurons suggests a cellular mechanism for learning. *Science,* 1983; 219:405-408.

Watkins, L.R. and Mayer, D.J. Organization of endogenous opiate and nonopiate pain control systems. *Science,* 1982; 216:1185-1192.

Whitehouse, W.G., Walker, J., Margules, D.L., and Bersh, P.J. Opiate antagonists overcome the learned helplessness effect but impair competent escape performance. *Physiology and Behavior,* 1983; 30:731-734.

Whitlow, J.W. Short-term memory in habituation and dishabituation. *Journal of Experimental Psychology: Animal Behavioral Processes*, 1975; 1:196-209.

Whitlow, J.W. and Wagner, A.R. Memory and habituation. In H.V.S. Peeke and L. F. Petrinovich (Eds.), *Habituation, sensitization, and behavior* (pp. 103-153). New York: Academic Press, 1984.

Willis, W.D., Sluka, K.A., Rees,H., and Westlund, K.N. Cooperative mechanisms of neurotransmitter action in central nervous sensitization. *Progress in Brain Research,* 1996: 110:151-166.

Wolpaw, J.R., Braitman, D.J., and Seegal, R.F. Adaptive plasticity in the primate spinal stretch reflex: initial development. *Journal of Neurophysiology*, 1983; 50:1296-1311.

Wolpaw, J.R., and Carp, J.S. Memory traces in spinal cord. *Trends in Neuroscience,* 1990: 13:137-142.

Wolpaw, J.R., and Lee, C.L. Memory traces in primate spinal cord produced by operant conditioning of H-reflex. *Journal of Neurophysiology*, 1989; 61:563 572.

Woolf, C.J. Evidence for a central component of post-injury pain hypersensitivity. *Nature,* 1983; 305: 686-688.

Woolf, C.J. Long term alterations in the excitability of the flexion reflex produced by peripheral tissue injury in chronic decerebrate rat. *Pain,* 1984; 18:325-343.

Woolf, C.J., and Thompson, S.W.N. The induction and maintenance of central sensitization is dependent on N-methyl-D-aspartic acid receptor activation; implications for the treatment of post-injury pain hypersensitivity states. *Pain,* 1991; 44:293-299.

3 PAVLOVIAN CONDITIONING OF FLEXION REFLEX POTENTIATION IN SPINAL CAT: TEMPORAL EFFECTS FOLLOWING SPINAL TRANSECTION

Russell G. Durkovic
Department of Neuroscience and Physiology
SUNY Health Science Center
Syracuse, NY 13210

3.1 INTRODUCTION

A long-lasting interest in basic neural mechanisms underlying the phenomena of learning and memory has produced a plethora of tactical approaches to discover and characterize them. Investigations have ranged from work on simple behaviors in invertebrate preparations (Goelet, Castellucci, Schacher and Kandel, 1986; Krasne and Glanzman, 1995)) to studies of neural plasticity phenomena in mammalian brain slices (Schwartzkroin and Wester, 1975; Okada and Miyamoto, 1989; Barrinuevo and Brown, 1983) and multielectrode recordings from intact animals (Mehta, Barnes and McNaughton, 1997). Another approach, incorporating the idea of relative neural simplicity but using the mammalian nervous system, has been the use of the spinal cord in classical and instrumental conditioning experiments (Patterson, 1975; Durkovic, 1975; Grau, Barstow and Joynes, 1998).

Two issues characterized the first few decades of work on such 'spinal conditioning' models. The first issue concerned doubts or confusion over how alterations in reflex activity in the spinal cord could be considered in the same context as learning and memory. As commonly considered by psychologists at the time such activities were thought to require important conscious components that were difficult to envision for the spinal cord. It was not until the realization that there were at least two major kinds of learning and memory (Tulving, 1987) (e.g., reflexive and declarative) that spinal conditioning began to gain general acceptance as an applicable model of the reflexive (or procedural) type of learning and memory.

The second issue was whether spinal conditioning actually existed, as many early attempts to show its existence failed (Patterson, 1975). Interestingly, most of the reported failures were from experiments conducted in adult, **chronic** (long-term) spinal animals. On the other hand most reported successes were carried out using adult, **acute** (just after transection) spinal animals (e.g., Patterson, 1975). It is this second issue that is the primary concern of the present chapter. This is because while the conditioned responses in the acute spinal animal are well characterized (Durkovic, 1986), the question remains how the adult, chronic spinal preparation would react under the same circumstances. As is described below, the chronic spinal cat does not seem to exhibit the same conditioning vivacity that acute spinal cats do when presented with the same Pavlovian paradigm.

The remainder of this chapter is organized in the following way. First, a review of what is known about spinal conditioning in acute spinal cat from work derived primarily from the author's laboratory is presented. Next, the results of spinal conditioning studies on two week and three month chronic spinal cats will be described. The results of additional studies on the time course of changes in reflex activity are introduced as partial explanations for the apparent alterations in conditionability in the chronic spinal animals. Finally, the potential relevance of spinal conditioning to: (i) learning and memory, (ii) mechanisms underlying allodynia and hyperalgesia and (iii) locomotor recovery from spinal injury is then discussed.

3.2 CLASSICALLY CONDITIONED FLEXION REFLEX FACILITATION IN ACUTE SPINAL CAT: ITS PARALLELS WITH CONDITIONED RESPONDING IN INTACT ANIMALS

Basic procedures

Male and female adult cats anesthetized initially with ether, tracheotomized, and respired with 1-3 % halothane in a 2:1 mixture of nitrous oxide and oxygen were placed in a rigid frame that fixed the head, hips and femur of the left hind limb (Durkovic, 1985). Following T-10 spinal transection and bilateral carotid artery ligation the animal was made decerebrate by clamping the vertebral arteries. Anesthetic was discontinued and the animal respired with room air for the remainder of the experiment. The distal tendon of the left tibialis anterior (TA) muscle was cut and attached to a strain gauge for measuring flexion reflex tension produced by electrical stimulation of saphenous (**S**) and superficial peroneal (**SP**) cutaneous nerves. While the recordings from TA are the principle responses to be described here, in other experiments recordings were also made from additional flexor muscles (semitendinosus, extensor digitorum longus). Recording electrodes attached to the cutaneous nerves allowed continuous monitoring of the nerve components activated and assurance of their constancy over the course of the experiment (Fig. 3.1).

Two hours after spinal transection trains of **S** stimuli (0.2 msec square wave pulses at 10/sec applied for 1.5 sec) were presented at one minute intervals to obtain preconditioning baseline flexion reflex response levels. The same train of (S) stimuli was then used as the conditioned stimulus (**CS**) in a Pavlovian conditioning paradigm with one group of animals receiving the **CS** paired with the **US** (the unconditioned stimulus: a train of 30/sec, 0.2 sec pulses delivered to the superficial peroneal nerve for 0.5 sec and overlapping the last 0.5 sec of the **CS** train). An unpaired control group received the same number of **CSs** and **USs** with the **USs** presented in the middle of the two minute interval between **CS** presentations. Stimulus intensity was adjusted to activate large myelinated A-beta and small myelinated A-delta fibers of each cutaneous nerve supramaximally, but was below the threshold for activating unmyelinated C fibers. Thirty conditioning trials were followed by thirty **CS-alone** trials at five minute intervals to obtain a measure of retention.

Pavlovian conditioning results in long-lasting flexion reflex potentiation

In a group of animals receiving **paired** trials the response to the **CS** increased over trials and reached asymptotic levels during conditioning (Figs. 3.2 and 3.3). In contrast the responses to the **CS** in the **unpaired** group declined somewhat in magnitude and differed significantly from the paired group.

During the retention trials following conditioning the responses to the **CSs** remained potentiated in the paired group and significantly elevated above that of the unpaired group for the retention period (over 2.5 hours) with no indication that the difference was changing over time (Figs. 3.2 and 3.3). If CS-alone trials following conditioning are presented at shorter (1 min.) intertrial intervals the reflex potentiation tends to be extinguished (Durkovic, 1975).

The arrow in each case represents the magnitude of the response to the initial CS-alone baseline trial. The length of the dotted line indicating CS application represents 1.5 sec. From Durkovic (1985) with permission.

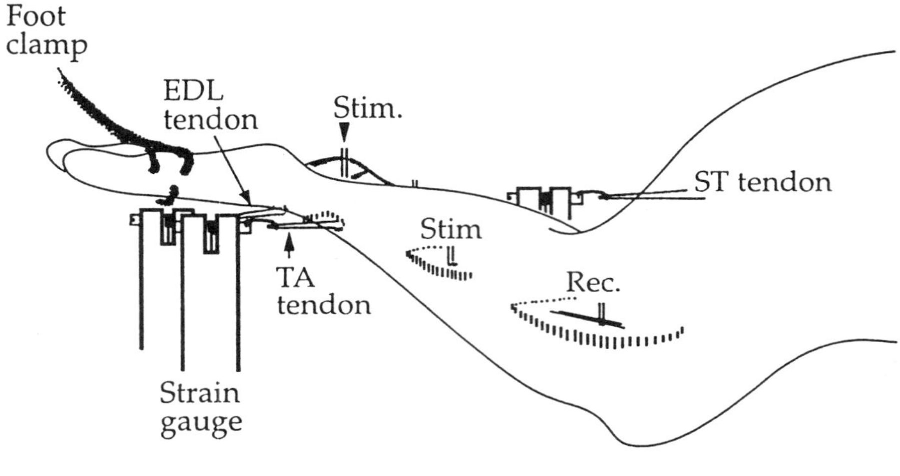

Figure 3.1. Drawing of a medial view of the left hind limb of a cat in an experimental setup.

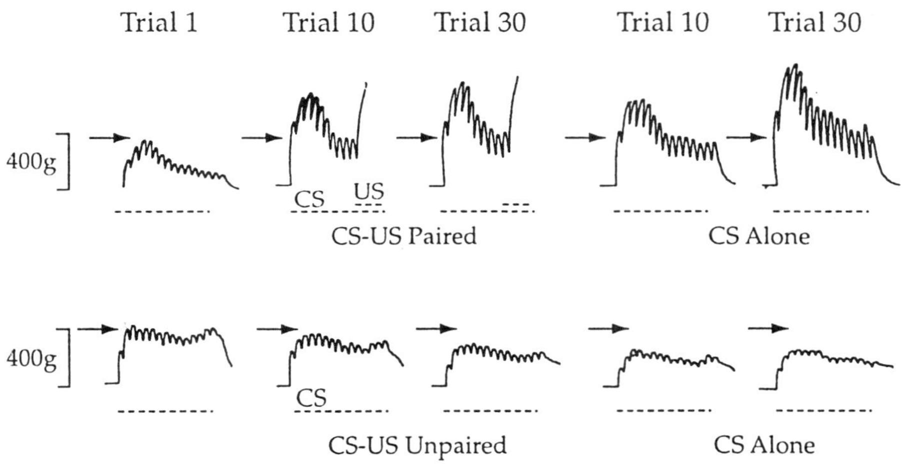

Figure 3.2. Representative TA muscle tension records from (A): an animal given paired CS-US presentations and (B): an animal given unpaired CS and US trials.

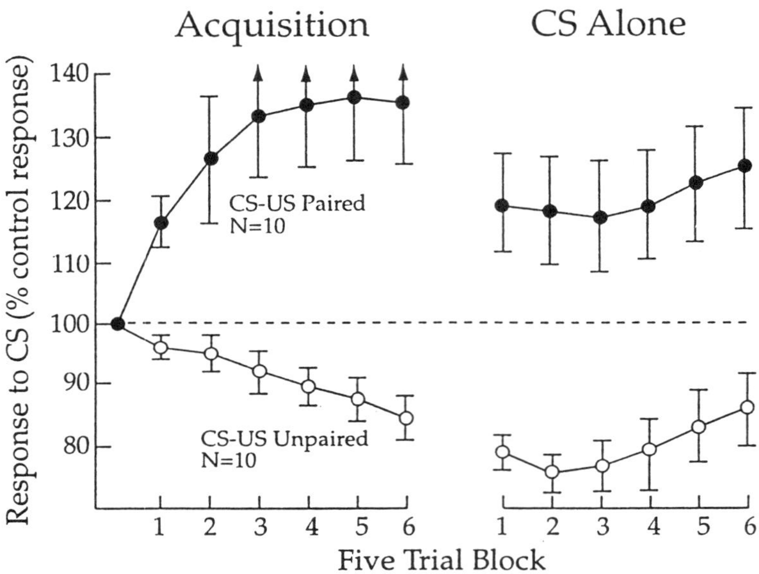

Figure 3.3. Group data for paired and unpaired animals' change from initial baseline control levels during conditioning (Acquisition) and CS-alone (Retention) trials. Control tensions justunder 600 grams for each group. From (Durkovic, 1985) with permission.

Interstimulus interval effects have significant influences on conditioned responding

Another study was undertaken to examine the effects of altering the interval between **CS** and **US** onset. Nine groups of animals with different interstimulus intervals ranging from -3 to +3 seconds were contrasted with an unpaired control group (Durkovic, 1986). When **CS** onset preceded **US** onset (forward conditioning) by 0.5 and 1.0 seconds there was significant facilitation compared to the unpaired control group (Fig. 3.4). However, when **CS** and **US** onsets occurred simultaneously or when **CS** onset preceded **US** onset by 0.25 or 3.0 seconds there were no significant differences from the unpaired control group.

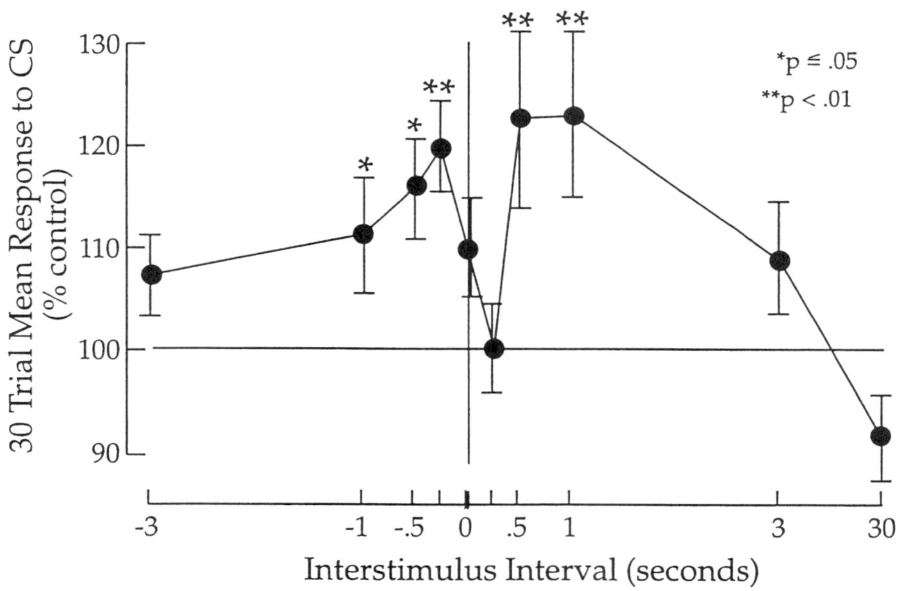

Figure 3.4. Interstimulus interval function curve for spinal conditioning. Redrawn fromDurkovic, 1996 with permission.

Surprisingly, when the **US** *preceded* the **CS** by 0.25, 0.5, and 1.0 second there was also significant reflex potentiation (i.e., 'backward conditioning' exists in this preparation). However, just as Pavlov's colleagues found in intact animals (Pavlov, 1927; Pavlov, 1928), the backward conditioned response is an effervescent one as during a 1/2 hour period of **CS**-alone trials following conditioning, only the 1.0 second forward conditioning group showed evidence of continued reflex enhancement compared to the unpaired control group. Another major difference between forward and backward conditioning in this preparation is that the potentiation occurs over different spinal reflex pathways. That is, the potentiation in the backward conditioning groups occurs over reflex pathways activated by A-beta S nerves (Hoover, 1989), while the potentiation in the forward conditioning groups occurs over reflex pathways activated by A-delta **S** nerves (Durkovic, 1983). These results suggest that forward and backward conditioning in this preparation may represent fundamentally different phenomena, as there are important behavioral and neurophysiological differences between the two.

Just why there is no potentiation of **A-beta** reflex pathways when there is complete overlap of CS and US pulses (i.e., when interstimulus intervals are 0.0 to 1.0 seconds) is not known. However, one possible explanation is the finding that there is a significant presynaptic inhibitory effect from **S** inputs that inhibit **SP** terminals under such conditions (Durkovic, 1995). This mechanism could prevent necessary associative excitatory inputs from the two nerves to the first order interneurons. Such dual excitatory effects may be required for development of potentiation within the neuronal circuits activated by **A-beta** cutaneous fibers.

CS and US intensity effects are analogous to those observed in intact animals

Studies of **CS** intensity effects for forward conditioning revealed an inverted U-shaped function curve (Durkovic, 1986, Fig. 3.4). That is, as **CS** intensity was varied in different groups from activation of A-beta **S** fibers (lowest intensity) to activation of A-beta and A-delta **S** fibers (middle intensity) to activation of all A and C **S** fibers (highest intensity) the only intensity resulting in significant differences between paired and unpaired control animals was the middle one. Intact animal conditioning studies show this same inverted U-shaped function curve when CS intensity is varied.

In contrast, forward spinal conditioning varies directly with US intensity as observed in most intact animal conditioning studies (A-beta **SP** fiber activation does not result in reflex potentiation while adding A-delta and then C **SP** fiber activation results in a concomitant enhancement of reflex potentiation [Durkovic, RG, 1986]).

Conclusion: Results of spinal conditioning studies show major parallels to classically conditioned responses of intact animals

As described above these studies suggest that many of the major characteristics that define classically conditioned responding in intact animal studies hold equally well for the flexion reflex alterations observed from the spinal cord in response to a Pavlovian conditioning paradigm. Parallels include:

(i) Response magnitudes increase over trials and reach plateau levels.
(ii) The conditioned reflex potentiation is long lasting.
(iii) The conditioned potentiation can be extinguished with short interval CS-alone trials.
(iv) The most efficacious reflex potentiation is a result of forward pairing and temporal contiguity of CS and US.
(v) Backward conditioning exists but is more effervescent than forward conditioning. Backward conditioning in this preparation may

represent a different phenomenon than forward conditioning since it involves different cutaneous nerve components for its elicitation.

(vi) Alterations in CS or US intensities produce conditioning changes consistent with results from intact animal conditioning studies.

(vii) The ability to exhibit differential conditioning in this preparation has been demonstrated (Beggs, 1985).

(viii) Simultaneous recordings from three different hind limb flexor muscles show that not all flexors exhibit conditioned reflex potentiation (Durkovic, 1983). As might be expected from intact animal studies, conditioned potentiation occurs only in those neuronal circuits displaying excitatory effects from both CS and US.

(ix) Higher order conditioned alterations that have been observed in intact animal conditioning studies have also been demonstrated in the spinal animal (Illich, Salinas and Grau, 1994)).

It should be noted that included in the above list are the principle characteristics identified by Neal Miller as defining "Grade-A Certified Learning" in **intact** animals (Miller, 1967).

3.3 RELATED EXPERIMENTAL RESULTS: FLEXION REFLEX CIRCUITRY, *C-FOS* ACTIVATION AND NMDA RECEPTOR BLOCKADE

A major finding in the circuitry of the flexion reflex was that the reflex is minimally disynaptic. That is, only a single interneuron need be interposed between cutaneous afferent terminals and flexor motoneurons (e.g., Leahy and Durkovic, 1991). Furthermore, a significant proportion of flexor motoneurons receive inputs via such a pathway. Additional studies employing anatomical markers pointed to spinal cord laminae V-VII and X locations for many of the interneurons involved in the disynaptic flexion reflex (Durkovic, Hoover, Light and Taylor-Blake, 1996). Furthermore, neurons located laminae I, II and IV are considered as additional potential locations of such interneurons. These results provide important insights into the neural network organization of the spinal circuits potentially involved in the conditioned reflex potentiation described above.

Because A-delta cutaneous fibers are known to project primarily to spinal cord laminae I, II and V (and A-delta cutaneous inputs are required for **forward** conditioning) it was hypothesized that activation of neurons in these three laminae would be of particular importance in the spinal conditioning phenomenon. To test this hypothesis the neuronal activity marker *c-fos* was used to investigate the loci of neuronal activation during spinal conditioning procedures (Nguyen and Durkovic, 1997).

Indeed, neurons expressing *c-fos* were more numerous in laminae I and II in animals receiving paired stimuli compared to those receiving unpaired stimuli. However, there were no significant differences in cell counts of *c-fos* labeled neurons in other laminae. These results suggest primary involvement of neurons in the two most dorsal laminae of the spinal cord in the long-term enhancement of A delta cutaneous reflex circuits brought about by the Pavlovian conditioning paradigm.

Central among proposed cellular mechanisms underlying learning and memory is the cascade of events responsible for long-term potentiation (LTP) in area CA-1 of the hippocampus (Durkovic, 1996). A major trigger for this cascade is glutamate activation of N-methyl-d-aspartate (NMDA) receptors. Interestingly, NMDA receptors are most heavily concentrated in spinal cord laminae I and II (Mitchell and Anderson, 1991), the region of the cord highlighted by the *c-fos* studies. Thus, experiments were conducted to examine the possible role of NMDA receptor activation in forward conditioned reflex potentiation (Durkovic and Prokowich, 1998). Animals were perfused intrathecally throughout the experiment with either artificial cerebrospinal fluid (CSF) or CSF containing 0.5 mM APV, an NMDA receptor blocker. During the 30 trial conditioning period there was no significant difference in the amount of A-delta flexion reflex enhancement between the two groups (Fig. 3.5, CONDITIONING).

However, during the 2 1/2 hour retention test period, significant differences between the two groups emerged as the APV group's response to A-delta S inputs fell below preconditioning baseline levels in contrast to the CSF group (Fig. 3.5, RETENTION). These results suggest that activation of NMDA receptors plays an important role in the induction of associative long-term potentiation of the flexion reflex and, furthermore, implies that certain cellular mechanisms of neural plasticity at the spinal level share the same properties as those utilized in higher neural centers.

3.4 RESULTS OF PAVLOVIAN CONDITIONING EXPERIMENTS IN CHRONIC SPINAL CAT

Experiments the same as those described above on acute spinal cats were carried out in cats with T-10 spinal cord transections made either 2 weeks or 3 months before this final experimental procedure. The only other difference in experimental protocol was the use of a one minute intertrial interval for both conditioning and retention periods. The assumption at the time was that with more time to recover from the trauma of the surgical transection, chronic spinal animals would exhibit even better conditioned responding than found a few hours after transection. The results were the just the opposite of what was expected.

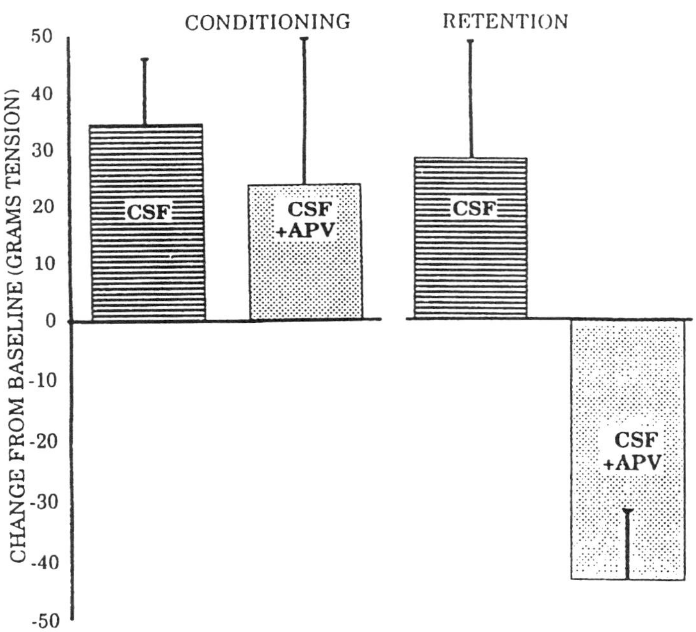

Figure 3.5. Change from baseline reflex response to A-delta cutaneous nerve inputs for control (CSF) and APV perfused animals. Thirty-trial mean response change (+SEM) during conditioning (left) and retention (right). Reprinted from (Durkovic, 1998) with permission from Elsevier Science.

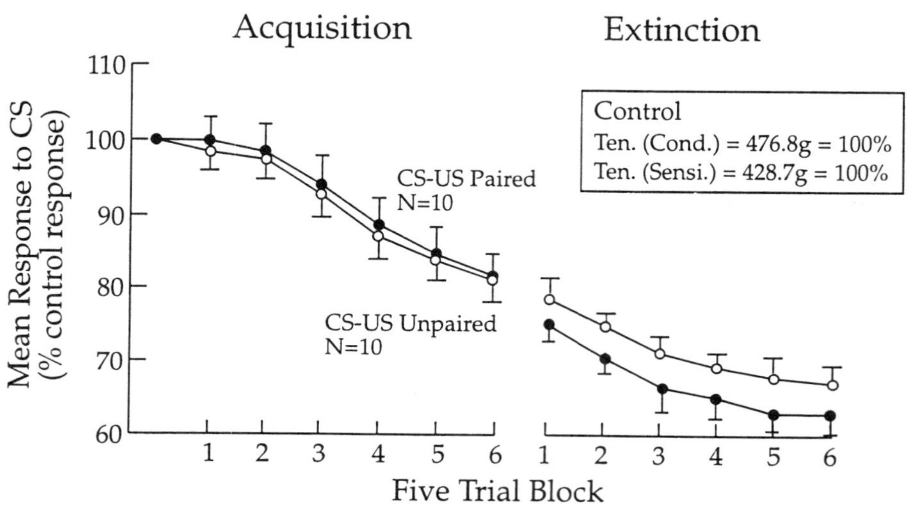

Figure 3.6. Change in TA response to the CS for two week chronic spinal animals receiving paired (filled circles) and unpaired (open circles) trials during conditioning and extinction (CS-alone) trials. Error bars are S.E.M.

In the case of the 2 week chronic spinal animals the reflex response to **S** inputs declined gradually over trials in both paired and unpaired (10 animals in each group) stimulus groups (Fig. 3.6). No significant differences in flexion reflex response magnitudes were observed between these groups for either the conditioning or the following CS-alone periods for these animals.

For three month chronic spinal animals a modest difference in reflex response between paired and unpaired animals (10 animals in each group) was observed during the 30 minute conditioning period. The responses of the paired group, while not showing much potentiation above initial baseline levels, were significantly greater than responses from the unpaired control group (p< 0.5). During the 30 minute CS-alone period following conditioning however, flexion reflex response magnitudes were virtually identical for the two groups (Fig. 3.7).

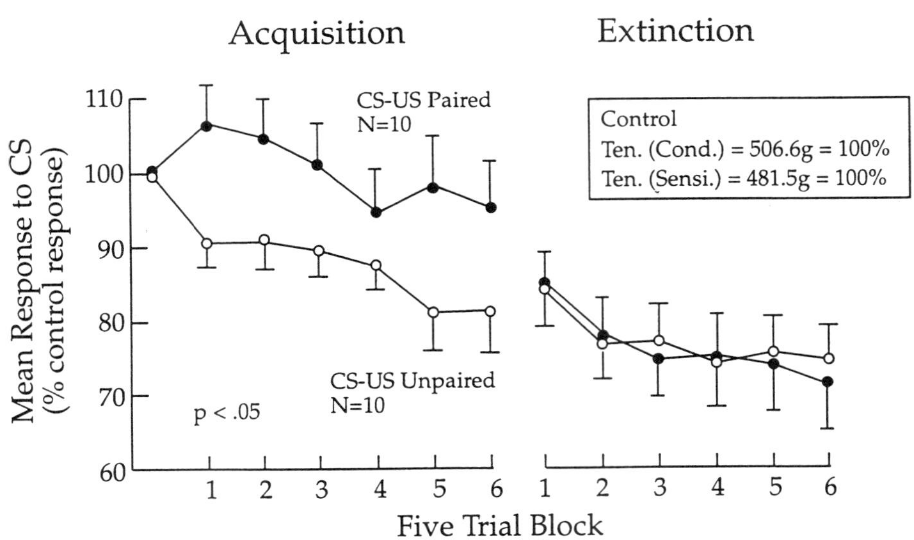

Figure 3.7. Change in TA response to the CS for three month chronic spinal animals receiving paired (filled circles) and unpaired (open circles) trials during conditioning and extinction (CS-alone) trials. Error bars are S.E.M.

3.5 ALTERATIONS IN RECRUITMENT ORDER AND IN FLEXION REFLEX EXCITABILITY FOLLOWING SPINAL TRANSECTION HELP EXPLAIN THE CONDITIONING RESULTS IN CHRONIC SPINAL ANIMALS

One of the basic physiological principles of muscle activation is that motor units are activated relative to their size, with small motor units being activated initially and larger ones activated when greater muscle forces are produced. This principle, known as the "size principle of motor unit recruitment " (Hennemen, 1977), has been shown to apply for the vast majority of movements including the flexion reflex (Durkovic, 1995). Furthermore, recruitment of larger motor units (using constant stimulus intensities) as conditioning trials progress is a mechanism involved in the conditioned reflex potentiation (Misulis and Durkovic, 1982). Recruitment order in the flexion reflex has been studied in acute and two week and three month chronic spinal animals by teasing the distal cut end of the nerve innervating the TA muscle and recording from a small fascicle of the nerve while stimulating **S** and **SP** at ever-increasing stimulus intensities (Durkovic, 1995).

Recordings from **acute** spinal cats showed that a majority of unit pairs exhibited 'orderly recruitment' in that small nerve action potentials were activated at lower cutaneous nerve stimulus intensities and larger action potentials were activated at higher intensities. This finding held true for both **S** and **SP** nerve stimuli (Fig. 3.8, A&B). On the other hand very different results were obtained **for two week chronic** animals. In these animals the **larger** of two action potentials was activated at a lower stimulus intensity more often when either cutaneous nerve was stimulated (Fig. 3.9, A&B). **Three month chronic** spinal animals were very similar to acute spinal animals, showing orderly recruitment (Fig. 3.10, A & B).

Another set of experiments compared tibialis anterior muscle tension evoked reflexively (10 Hz **S** stimuli) with muscle tension obtained by stimulating the motor nerve to this muscle at the same stimulus rate. For acute spinal cats the reflex/direct stimulus tension ratio was about 0.45. In two week and three month chronic spinal animals the ratios were about 0.95 and 0.60 respectively.

These results suggest that chronic spinal animals undergo a flexion reflex alteration compared to acute spinal animals for the first few weeks after spinal transection. Chronic spinal animals have a heightened flexion reflex excitability that would have the effect of almost totally (two week chronics) or partially (three month chronics) limiting the magnitude of the reflex enhancement that could be brought about by Pavlovian conditioning techniques. Furthermore, an altered recruitment order in two week chronic spinal cats is consistent with the near total motoneuron recruitment by the **S** flexion reflex, exemplifying the ease by which large motoneurons may be activated by cutaneous nerve inputs in these animals. The results further suggest that there may be a gradual return toward the reflex excitabilities found in the acute spinal cat. Concomitantly, a return of conditioned reflex capabilities may take place gradually over time following transection as the

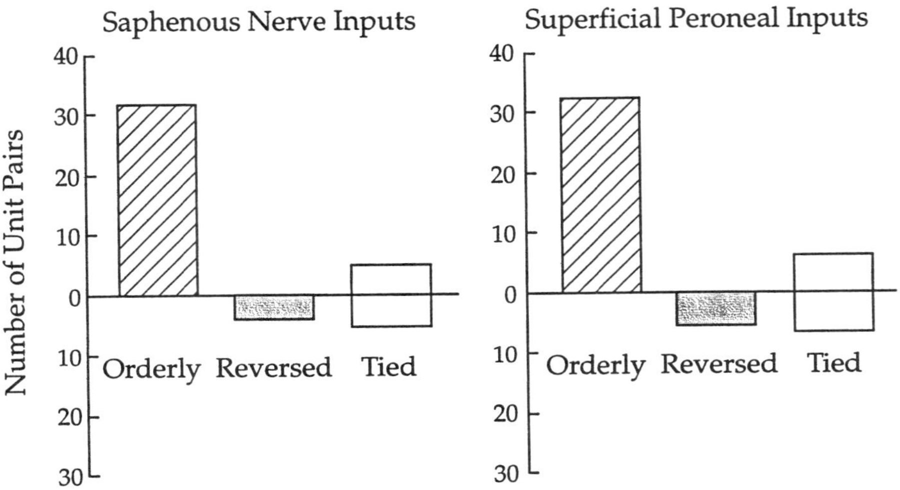

Figure 3.8. Recruitment order for TA nerve axon pairs in acute spinal animals.

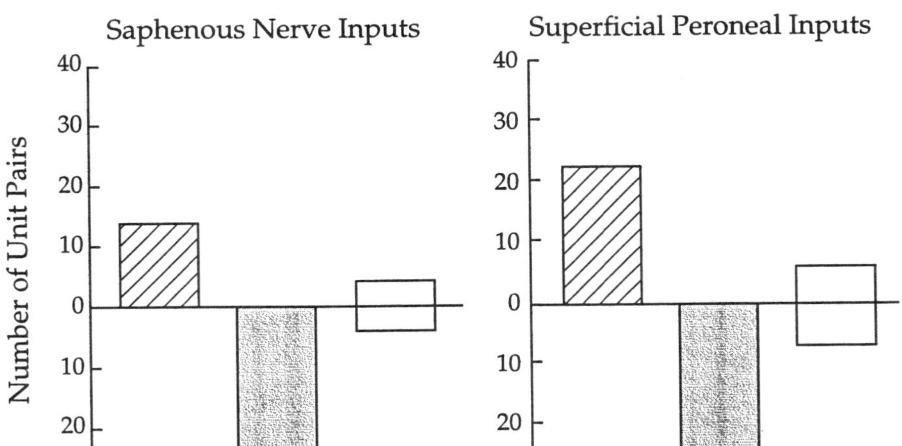

Figure 3.9. Recruitment order for TA nerve axon pairs in two week chronic spinal animals showing a tendency for recruitment order reversal.

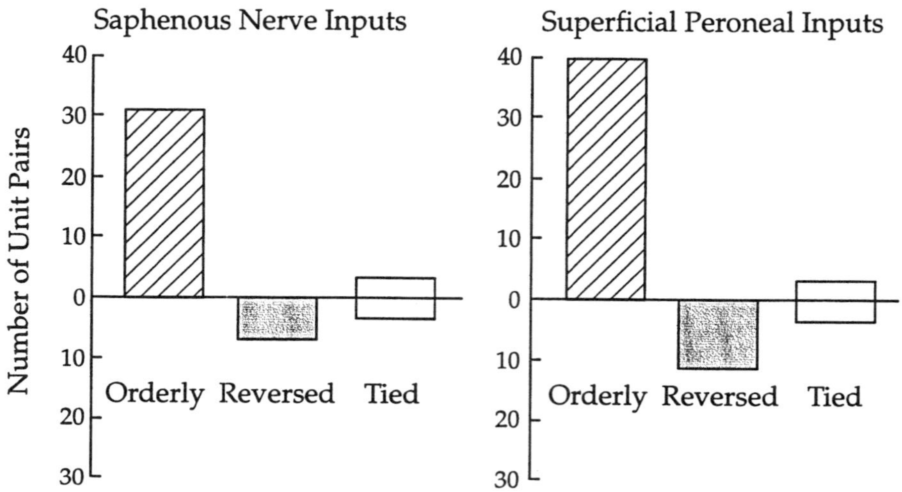

Figure 3.10. Recruitment order for TA nerve axon pairs in three month chronic spinal animals.

three month chronic spinal animals were able to exhibit some conditioned flexion reflex enhancement during conditioning.

3.6 POTENTIAL RELEVANCE OF SPINAL CONDITIONING TO MECHANISMS OF LEARNING AND MEMORY, ALLODYNIA AND HYPERALGESIA, AND TO RECOVERY OF LOCOMOTOR BEHAVIOR FOLLOWING SPINAL INJURY

As a simplified neural model of reflexive learning, 'spinal conditioning' exhibits the principle characteristics that have been ascribed to this form of learning and memory in intact animals (see above). While not simple, the neural model is simplified compared to the intact animal and thus offers great potential in helping to uncover cellular neurophysiological mechanisms underlying mammalian learning and memory. The apparent existence of two unique forms of learning and memory at the spinal level (those of forward and backward conditioning) is just one example of how this preparation may contribute to an understanding of these processes.

Furthermore, insights into the nervous system alterations producing the pain syndromes of allodynia or hyperalgesia might also be revealed from analyses of the mechanisms underlying these two spinal conditioning phenomena. At the spinal level each of these clinical problems might be expected to alter reflex function in much the same way observed in the conditioning studies. This is because a direct relationship has been observed between pain sensation and the magnitude of certain flexion reflexes in human (Willer, 1977). Thus allodynia, a sensation of pain in response to previously non-painful inputs, might be modeled by the enhanced reflex output to activation of A-beta fibers, stimulation of which are generally not associated with painful sensation. Given that backward conditioning produces enhancement of reflex circuits activated by A-beta fibers, the same cellular mechanisms that are involved in backward conditioning may be involved in the generation of allodynia as well. On the other hand hyperalgesia, a more painful sensation than normal to a previously painful input, might be modeled by forward conditioning mechanisms in the spinal cord. This is because forward conditioning is localized to circuits activated by A-delta fibers, many of which carry painful sensory information. Enhanced flexion responses to these inputs might be expected in individuals experiencing hyperalgesia.

Finally, because last order interneurons of the flexion reflex appear to be a component of the central pattern generator for locomotion (Andersson, Forssberg, Grillner and Lindquist, 1978), it is possible that conditioned reflex potentiation procedures could prove useful in enhancing locomotor capabilities in spinal injured individuals. In chronic spinal cat other forms of training-induced enhancement of locomotion have proved to be successful (deLeon, Hodgson, Roy and Edgerton, 1998) and these might be combined with Pavlovian paradigms in spinal injured humans to generate the greatest amount of recovery of function.

3.7 SUMMARY

In acute spinal cats a Pavlovian conditioning paradigm (paired trains of stimuli applied to two different cutaneous nerves) produces a potentiation of flexion reflexes. The characteristics of this potentiation match the characteristics of classically conditioned responses that have been observed in intact animals. This associative flexion reflex LTP requires NMDA receptor activation and is associated with enhanced activity of laminae I and II spinal interneurons.

The same paradigm presented to two week chronic spinal cats does not result in reflex potentiation and no differences from control animals given unpaired stimuli are observed. Three month chronic spinal cats recover a portion of the conditioning effect that is observed in acute spinal cats.

Additional studies indicate that two week chronic animals exhibit abnormal motoneuron recruitment orders compared to the other two groups. This recruitment order change is related to an increased baseline flexion reflex excitability when cutaneous nerves are activated. With time, the baseline reflex excitability declines back toward the levels that are found following acute transection. These excitability changes appear to prevent the observations of the motor manifestations of classically conditioned reflex potentiation in two week chronic spinal animals and limit the potential for such conditioning in three month chronic animals. However, the return of some conditioning capability in the three month chronic spinal animals suggests that over time, adult spinal animals may regain the capability of exhibiting classically conditioned flexion reflex potentiation.

3.8 REFERENCES

Andersson, O., Forssberg, H., Grillner, S., Lindquist, M. Phasic gain control of the transmission in cutaneous reflex pathways to motoneurons during 'fictive' locomotion. *Brain Research,* 1978; 149: 503-507.

Barrionuevo, G, Brown, T.H. Associative long-term potentiation in hippocampal slices. *Proceedings of the National Academy of Sciences USA,* 1983; 80: 7347-7351.

Beggs, A.L., Steinmetz, J.E., Patterson, M.M. Classical conditioning of a flexor nerve response in spinal cats: effects of tibial nerve CS and a differential conditioning paradigm *Behavioral Neuroscience,* 1985; 99:496-508.

de Leon, R,, Hodgson, J.A., Roy, R.R., Edgerton, V.R. Locomotor capacity attributable to step training versus spontaneous recovery following spinalization in cats. *Journal of Neurophysiology,* 1998; 79:1329-1340.

Durkovic, R.G. Classical conditioning, sensitization and habituation in spinal cat. *Physiology and Behavior,* 1975; 14:297-304.

Durkovic, R.G. Classical conditioning of the flexion reflex in spinal cat: features of the reflex circuitry. *Neuroscience Letters,* 1983; 39:155-160.

Durkovic, R.G. Retention of a classically conditioned reflex response in spinal cat. *Behavioral and Neural Biology,* 1985; 43:12-20.

Durkovic, R.G. "The spinal cord: a simplified system for the study of the neural mechanisms of learning and memory." In *Development and Plasticity of the Mammalian Spinal Cord*, Michael E.Goldberger, Alfredo Gorio, Marion Murray, eds. New York, NY: Springer Verlag, 1986.

Durkovic, R.G. Presynaptic inhibition: a test of its role in blocking classically conditioned reflex potentiation at certain interstimulus intervals. *Society for Neuroscience Abstracts,* 1995; 21: 389.

Durkovic, R.G. "Plasticity of motoneurone recruitment order in the flexion reflex following spinal cord transection." In *Alpha and Gamma Motor Systems,* Anthony Taylor, Margaret H. Gladden, Rade Durbaba, eds. New York, NY: Plenum Press, 1995.

Durkovic, R.G. "Hippocampal Long-Term Potentiation: A Model of the Cellular Processes Underlying Learning and Memory." In: *The Scientific Foundations of Neurology*. A.Norman Guthkelch and Karl E. Misulis, eds. Cambridge, MA: Blackwell Science, 1996, pp. 91-108.

Durkovic, R.G, Damianopoulos EN. Forward and backward classical conditioning of the flexion reflex in the spinal cat. *Journal of Neuroscience,* 1986; 6:2921-2925.

Durkovic, R.G, Hoover, J.E., Light, A.L., Taylor-Blake, B. Anatomy of the disynaptic flexion reflex circuitry in spinal cat. *Society for Neuroscience Abstracts,* 1996; 22:1835.

Durkovic, R.G., Prokowich, L.J. D-2-amino-5-phosphonovalerate, an NMDA receptor antagonist, blocks induction of associative long-term potentiation of the flexion reflex in spinal cat. *Neuroscience Letters,* 1998; 257:162-164.

Goelet, P., Castellucci, V.F., Schacher, S., Kandel, E.R. The long and the short of long-term memory – a molecular framework. *Nature,* 1986; 322:419-422.

Grau, J.W., Barstow, D.G., Joynes, R.L. Instrumental learning within the spinal cord: I. Behavioral properties. *Behavhavioral Neuroscience,* 1998; 112:1366-1386.

Hennemen, E. Functional organization of motoneuron pools: The size principle. *Proceedings of the International Union of Physiological Sciences,* 1977; 12:50.

Hoover, J.E., Durkovic, R.G. Retention of backward classically conditioned reflex response in spinal cat. *Experimental and Brain Research,* 1989; 77:621-627.

Illich, P.A., Salinas, J.A., Grau, J.W. Latent inhibition, overshadowing and blocking of a conditioned antinociceptive response in spinalized rats. *Behavioral and Neural Biology,* 1994; 62:140-150.

Krasne, F.B., Glanzman, D.L. What can we learn from invertebrate learning. *Annual Review of Psychology,* 1995; 46:585-624.

Leahy, J.C., Durkovic, R.G. Differential synaptic effects on physiological flexor hindlimb motoneurons from cutaneous inputs in spinal cat. *Journal of Neurophysiology,* 1991; 66: 460-472.

Mehta, M.R., Barnes, C.A., McNaughton, B.L. Experience-dependent, asymmetric expansion of hippocampal place fields. *Proceedings of the National Academy of Science,* 1997; 94:8918-8921.

Miller, N.E. " Certain facts of learning relevant to the search for its physical basis." In *The Neurosciences: A Study Program*, Gardner C. Quarton, Theodore Melnechuk, Fransis O. Schmitt, eds. New York, NY: Rockefeller University Press, 1967.

Misulis, K.E., Durkovic, R.G. Classically conditioned alterations in single motor unit activity in the spinal cat. *Behavior and Brain Research,* 1982; 5:311-317.

Mitchell, J.J., Anderson, K.J. Quantitative autoradiographic analysis of excitatory amino acid receptors in the cat spinal cord, *Neuroscience Letters,* 1991; 124:269-272.

Nguyen, M.L., Durkovic, R.G. Classically conditioned *c-fos* expression in spinal cat. *Society for Neuroscience Abstracts*, 1997; 22:1835.

Okada, Y., Miyamoto, T. Formation of long-term potentiation in superior colliculus. slices from the guinea pig. *Neuroscience Letters,* 1989; 96:108-113.

Patterson, M.M. "Mechanisms of classical conditioning and fixation in spinal mammals." In *Advances in Psychobiology,* Vol. III, Austin Riesen, Richard F Thompson (Eds.) New York, NY: Wiley, 1975.

Pavlov, Ivan P. *Conditioned Reflexes*. London: Oxford University Press, 1927, p. 393.

Pavlov, Ivan P. *Lectures on Conditioned Reflexes*. New York: International Publishers Company, 1928, p. 381.

Schwartzkroin, P.A., Wester, K. Long-lasting facilitation of a synaptic potential following tetanization in the *in vitro* hippocampal slice. *Brain Research,* 1975; 89:107-119.

Tulving, E. Multiple memory systems and consciousness. *Human Neurobiology,* 1987; 6:67-80.

Willer, J.C. Comparative study of perceived pain and nociceptive flexion reflex in man. *Pain,* 1977; 3:69-80.

4 SPINAL FIXATION: LONG-TERM ALTERATIONS IN SPINAL REFLEX EXCITABILITY WITH ALTERED OR SUSTAINED SENSORY INPUTS

Michael M. Patterson
Department of Osteopathic Principles and Practice
College of Osteopathic Medicine
NOVA Southeastern University
Ft. Lauderdale, Florida 33328

4.1 INTRODUCTION

The mammalian spinal cord has long been known to contain a rich variety of reflex pathways, in addition to providing the pathways for transmission of neural signals to and from the body and the higher nervous system. For many years, these reflexes were assumed to be essentially unchanging, hard-wired entities that provided the underpinnings for response to stimuli. Sherrington (1905) noted that reflex activity could be increased or decreased with activation, but that these effects were transient. In the 1960s and 1970s work by investigators such as Thompson (e.g.,Thompson and Spencer, 1966) began to study the ability of sensory inputs to the reflex pathways to alter reflex activity. Soon, the work of Kandel (e.g., Carew, Pinsker and Kandel, 1972) and his associates showed in *Aplysia* that reflex alterations could have a longer-lasting component. This led to the realization that in the mammalian cord there could be longer lasting aspect of reflex excitability alterations. However, this fact had been shown to occur with learning paradigms much earlier. Shurrager and his colleagues (e.g., Shurrager and Culler, 1940) had demonstrated that spinal reflexes responded with increased activity to a classical conditioning paradigm and that the increased responsiveness lasted for hours in spinalized animals. While controversial, these data set the stage for increased interest in the possibility that spinal reflexes were not simply relatively unchanging neural networks left over from the distant past and superceded by increasing encephalization of function in higher animals. The demonstrations that spinal reflexes could be altered by learning paradigms necessitated the corresponding assumption that spinal reflex pathways contained a type of memory mechanism in addition to the capacity to respond to stimulus contiguity. This memory function was presumably a mechanism that was precipitated by activation of the circuits and would last for some time after the stimulus ceased.

In fact, memory in the spinal cord had been demonstrated experimentally many years before in the laboratories of A.M. DiGiorgio (1929). DiGiorgio had shown that a cerebellar lesion in anesthetized animals could produce an asymmetrical posture in the hind limbs, with one flexed actively and the other extended. If the animal was left in this posture for several hours then the spinal cord severed at the mid back, the postural asymmetry would remain. The flexion of the limb was assumed to have been "fixated" in the spinal cord by the altered outflows from the cerebellar damage. This spinal fixation was one of the initial demonstrations of memory in the spinal cord.

4.2 THE PARADIGM'S BEGINNINGS

The work begun by DiGiorgio in 1929 led to other studies from her laboratories that showed essentially the same phenomenon in dogs and rabbits (DiGiorgio, 1943, 1947a, 1947b, 1947c). The cerebellar lesions would produce a postural asymmetry that persisted following spinal cord transection. This persistence of an active muscular contraction following spinal transection implied that the action of altered outflows from the brain injury acting over time could alter the excitability of the spinal reflex pathways sufficiently that abolishing the activating neural activity would not abolish the activity in the spinal cord. Soon others began investigating the effect. Giulio (1952) and Manni (1950) found the same effect in guinea pigs and pigeons. Alella (1948) studied the effect more extensively by making lesions of the sensory-motor and parietotemporal cortex. He produced flexion-extension patterns that after 90 minutes to several hours, persisted after cord transection. The patterns were seen to last for 2-4 days. He noted that a reversal of the asymmetry during hypothermia was seen in guinea pigs (for a full review, see Patterson, 1976).

In the early to mid 1960s, Ralph Gerard and his student Chamberlain performed a series of studies on the DiGiorgio persistence of spinal flexion. They found that in rats, the effect would persist in all animals if 45 minutes were allowed to elapse between successful brain lesion and spinal transection, but would not last with only a 35 minute wait (Chamberlain, 1961, 1963). They also showed that various lesions of the CNS would produce the effect (Chamberlain, Halick and

Gerard, 1963; Chamberlain, Rothschild and Gerard, 1963). They hypothesized that the effect was dependent on alterations in RNA synthesis or structure, and studied the effects of RNA synthesis inhibitors and enhancers on the fixation process. They found that inhibitors generally increased spinal fixation time (SFT) from 45 to 70 minutes, and enhancers tended to decrease it, from 45 to 25-30 minutes, in line with expectations. Thus, they felt that spinal fixation was due to alterations in RNA synthesis or cellular metabolism caused by alterations in activity on the neuron.

Following the Gerard studies, a number of investigators looked at the fixation paradigm in an effort to determine its physiological basis. Giurgea and his colleagues (Giurgea, 1973; Giurgea, Daliers and Rigaux, 1971; Giurgea and Mouravieff-Lesuisse, 1971; Giurgea, Mouravieff-Lesuisse and Daliers, 1973) looked at the effects of various drugs on the SFT in the rat model. They generally found that drugs enhancing performance in normal learning situations decreased SFT, while drugs impairing performance in these situations increased SFT. Some drugs, including stimulants, did not affect SFT. They concluded that spinal fixation was related to protein metabolism. In a separate study, Mouravieff-Lesuisse and Giurgea (1970) found that electroconvulsive shock (ECS) seemed to prolong SFT but that the drugs seen to shorten SFT could partially reverse the effects of ECS. Mouravieff-Lesuisse and Guirgea (1973) also found that cortical spreading depression (CSD) on the side ipsilateral to the cerebellar lesion prolonged SFT to about 70 minutes without affecting the asymmetrical posture, while contralateral CSD and saline controls showed no alteration in SFT.

In the 1960s, the possibility that memory could be influenced by brain extracts as propounded by Jacobson and his colleagues (Jacobson, Fried and Horowitz, 1966) led Guirgea and his colleagues to asses the effects of spinal cord extracts on SFT (Daliers and Giurgea, 1969; Giurgea et al., 1971). These studies utilized extracts from donor rats given cerebellar lesions and maintained for varying periods of time. An extract of the donor brains when injected I.P. into recipient rats prior to or at the time of cerebellar lesions was found to shorten SFT by about 10 minutes. The extract worked if the donor had sustained an asymmetry but not if no asymmetry had been present, and only of the recipient had an asymmetry-producing lesion. The authors concluded that the extract was an oligopeptidic material that could facilitate the fixation process, but not produce it. These studies led to a series of studies by Russian investigators (see below) showing much the same results 25 years later.

These early investigations showed a process that had memory-like features that took some time to consolidate and could be affected by various drugs and memory enhancers. They also showed that the process could be influenced by the cortex, but that cortical intervention was not necessary, as Chamberlain often used decerebrate animals in his studies. Protein involvement in the process was indicated, as shown by the drug studies.

Further studies on the fixation effect were performed by various groups during the late 1960s and early 1970s. Palmer and his colleagues (Palmer, Ward and Davenport, 1968; Palmer and Davenport, 1969; Palmer, Davenport and Ward, 1970; Palmer, Davenport and Ward, 1970; Palmer, Davenport and Ward, 1970; Palmer,

Davenport and Ward, 1970) used newborn rats in their studies. They made small stab wounds in the anterior cerebellum of 1-3 day old rat pups. The resulting contralateral hind limb extension was observed to be fixated in from 38 to 45 minutes. Metabolic inhibitors lengthened the SFT to 83 minutes while eserine and reserpine increased it to about 60 minutes. From their studies, they suggested that neural inhibition or neural fatigue from overexcitaiton from the cerebellar lesion could account for the effects seen. They found that labeled leucine was incorporated more slowly into the cord in fixated pups. They suggested that either neuronal inhibition or neuronal fatigue from overexcitation due to the cerebellar lesion could account for these findings. However, they found that labeled leucine was incorporated most rapidly about 75 minutes after it was injected in the lesioned animals but 60 minutes after injection in the control animals. From their findings, they suggested that RNA synthesis alterations were possibly the basis for the fixated excitation. They also looked at several other drugs, such as eserine, atrophine, adrenergic drugs, reserpine, and amphetamine. The results of these studies were very mixed, sometimes showing increases and sometimes decreases in SFT, depending on dose and time. The studies indicated that the effects of drugs on SFT were complex, varying with type of drug, dose, and time given. The immediate conclusions from the Palmer studies must be that drug effects on SFT should be interpreted cautiously and that parametric studies of drug effects are necessary for proper conclusions to be drawn.

Another series of studies showing a phenomenon akin to spinal fixation, but in humans was carried out in the 1940s and early 1950s by Denslow and Korr (Denslow, 1944; Denslow, Korr and Krems, 1947; Korr, 1947). These investigators and their colleagues found that in otherwise normal human subjects, the excitability of reflexes often showed long-lasting and individual variation. These areas of "chronic facilitation" were shown by lowered thresholds of elicitation of EMG activity in the back muscles and also by increased sympathetic activity in the area. The areas of low threshold were often activated by inputs at remote sites and by psychological stressors that would cause no activity in adjacent areas. The authors suggested that the chronic facilitation of motoneuron pools seen in the studies was most likely due to altered sensory input or ongoing proprioceptive inputs. The long-term and individual nature of the altered excitatory areas suggests that they are the result of alterations of spinal circuits rather than continued inputs from the periphery.

In 1947, Frankstein (1947) reported another type of spinal fixation. He injected turpentine into the footpads of cats and waited for the inflammation to recede. For several days following the injections the animals favored the paw and often held it in a flexed position. After the animals had recovered and no longer walked with a limp, Frankstein decerebrated the animals. Rather than the normal decerebrate rigidity with all limbs extended, the previously injected limb was flexed or flaccid. Frankstein interpreted this to indicate that while the peripheral injury had healed, long-lasting alterations in the spinal organization had been made. With the cortex intact, the animal had compensated for the spinal changes. With the cortex removed, the alterations in spinal excitability became evident. These studies are in line with the Denslow and Korr findings in humans in suggesting that spinal excitability can be altered by peripheral inputs.

The studies performed on the spinal fixation phenomenon from the first demonstration in 1929 through the 1970s suggested that alterations of descending influences could act on spinal circuits to alter the excitability of spinal neurons. The effects of drugs on the process was shown to be complex but that generally, drugs that inhibited memory lengthened SFT, while memory enhancing drugs seemed to shorten it. Some of the data indicated that a substance elaborated in animals undergoing spinal fixation could enhance the process in other animals. The process was viewed as a form of memory in the spinal cord, perhaps mediated by some alteration in RNA structure. The studies cited in humans suggested that a similar process could occur in humans, resulting in altered reflex thresholds in both somatic and sympathetic pathways.

4.3 REDEFINING THE PARADIGM

In the early 1980s my graduate students (primarily Joe Steinmetz) and I began a reevaluation of the spinal fixation phenomenon. We had been actively studying classical conditioning of spinal reflexes for several years (e.g., Patterson, 1980). Classical conditioning necessitates a form of memory or change in functional characteristics of a neural pathway and the spinal cord reflex systems had been shown to be able to support such alterations. The phenomenon of spinal fixation was a much simpler paradigm than classical conditioning, necessitating only one stimulus rather than the contiguity of two stimuli. The possibility that alterations in spinal reflex excitability could be accomplished by a single stimulus over time was appealing from the standpoint that a wide variety of parameters could easily be investigated in such a preparation. By this time, it was becoming evident that sensitization and habituation, as well as long-term habituation readily occurred in mammalian spinal circuits and the possibility of longer lasting alterations seemed high.

We began our studies by replicating as closely as possible the procedures used by Chamberlain in his studies (Steinmetz, Cervenka, Robinson, Romano and Patterson, 1981). Rats were anesthetized to surgical levels and an electrolytic lesion placed in the anterior cerebellum. The lesion had to produce a marked flexion in a hind limb to be used. Once a lesion had been made and flexion observed either 20 or 50 minutes were allowed to elapse prior to spinal transection at about the T7 spinal level. The animal was suspended in a manner that allowed direct measurement of the amount of flexion by weights put in a small paper cup tied to the toe of the flexed foot. The amount of weight necessary to pull the flexed leg down to the level of the normal leg was recorded as the amount of asymmetry in grams. If the flexed leg did not return to a flexed position after measurement, the asymmetry was recorded as zero. Thus, a quantifiable measure of the active fixated flexion was possible. We found that in the animals allowed 50 minutes of flexion, all showed retention of the flexion after spinal cord section, while almost no retained flexion was measured in the 20 minute animals (see Figure 4.1). This study affirmed the observations of Chamberlain that at least 45 minutes had to be allowed for fixation to occur. In another study in the same series, we allowed 25, 30, 35, 40 or 45 minutes to elapse between cerebellar lesion and spinal transection in separate groups of rats (see Figure 4.2). We saw a phasing in of retention of the flexion, indicating that at 35 minutes, some retention occurred, but that by 45 minutes, all animals showed some fixation. These studies replicated the basic findings of Chamberlain and Gerard regarding the temporal parameters of the fixation phenomenon.

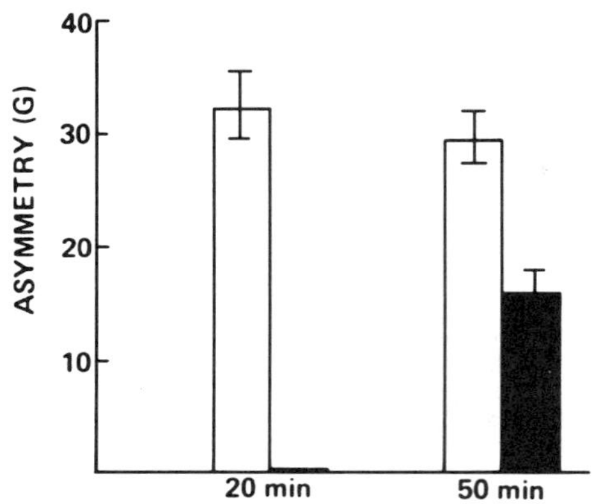

Figure 4.1. Mean amounts of presection asymmetry (open bars) and postsection asymmetry (black bars) present in groups allowed 20-min or 50-min spinal fixation time (ns = 15). Vertical lines = SE. Reprinted from Steinmetz, et al., 1981 with permission.

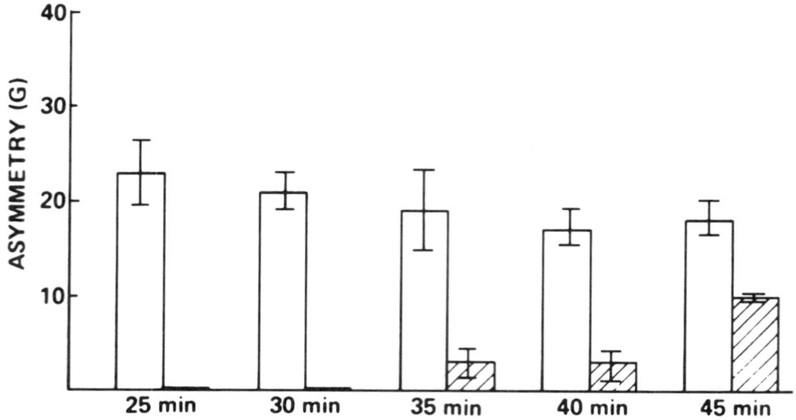

Figure 4.2. Mean amounts of presection asymmetry (open bars) and postsection asymmetry (cross-hatched bars) present in groups allowed spinal fixation time or 25, 30, 35, 40 or 45 min (ns = 5). Vertical lines = SE. Reprinted from Steinmetz, et al., 1981 with permission.

The earlier studies of spinal fixation had all used a central lesion of one sort or another as the initiator of the postural asymmetry that was fixated. There seemed to be an implicit assumption that the altered outflows from the injured brain area was necessary for the fixation process. The use of central lesions to produce the fixated postures was, however, inherently difficult. The necessity of stereotaxically placing lesions in a specific cerebellar area is both time consuming and somewhat error prone. The question arose as to what the necessary features were that produced the fixated postural alteration. Was the central lesion a necessary factor or was the flexed posture sufficient? If the central lesion was necessary to begin a process that resulted in a spinal memory formation, that would suggest some viable processes and would exclude others. If however, the central lesion was not mandatory, then the processes would be limited to the spinal cord itself and would be somewhat more circumscribed. We performed the third study in the series (Steinmetz, et al., 1981) to test the hypothesis that the central lesion was not a necessary feature of spinal fixation. Here, rats were anesthetized and prepared for a spinal transection at the T7 spinal level. They were then given either 20 or 45 minutes of 1-4 ma of 7 msec., 100 pps stimulation through stainless steel wound clips placed on the upper right hind thigh. This stimulus was sufficient to produce a tight flexion of the right leg. After stimulus termination, the right leg was measured for remaining flexion in the manner described above for the cerebellar lesion animals. A spinal transection was then performed and after 5 minutes, the flexion remaining was again measured. The results are shown in Figure 4.3. The rats stimulated for 20 minutes showed the same amount of remaining flexion after stimulus termination as did the 45-minute group, but almost no remaining flexion after spinal transection, whereas the 4-minute group showed a robust retention of flexion after spinal transection. This study seemed to show that spinal fixation could result from peripheral stimulation in the absence of a central lesion (a finding in concert with Frankstein's data).

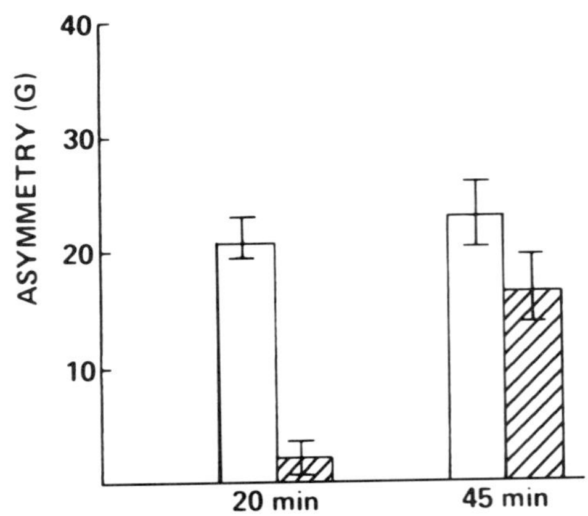

Figure 4.3. Mean amounts of presection asymmetry (open bars) and postsection asymmetry (cross-hatched bars) present in groups allowed 20-min or 45-min hind limb stimulation (ns = 7) Vertical lines = SE. Reprinted from Steinmetz, et al., 1981 with permission.

4.4 DEFINING THE PREPARATION

With the basic proof that spinal fixation could be accomplished without a central lesion, but by a peripheral stimulation to produce the postural asymmetry, we proceeded to look at various aspects of the preparation. In the same series as above, we determined that the peripheral stimulation was not having its effect directly on muscle by activating muscle fibers and causing contracture. We gave separate groups of animals complete dorsal and ventral rhizotomies from T10-L6, thus severing all nerves to and from the spinal cord and the hind limb. Five groups of animals were stimulated as above for 10, 20, 30, 40 or 50 minutes. The stimulus produced a tight flexion in the right hind limb for the time indicated. Remaining flexion was measured immediately and 2 hours after stimulation. No spinal transection was performed. Remaining asymmetry was almost none, with only the 40 and 50 minute groups showing flexion above 2.5 grams (as opposed to 15-20 grams in the nonrhizotomized animals) immediately and only one animal showing any after 2 hours. Thus, the disconnection of the sensory and motor nerves from the hind limb resulted in essentially no retained postural asymmetry. It appeared that the spinal cord was necessary for the retained flexion.

Was the brain itself necessary for the fixation of the altered excitability in the spinal cord? To answer this, we gave two groups of animals spinal transections prior to the hind limb stimulation of 20 and 45 minutes. The results showed that the 20-minute group had about 5 grams of retained flexion after the stimulus was terminated, whereas the 45-minute group had about 18 grams. This figure was almost identical to the 45-minute group given stimulation with an intact cord and measured after stimulation and spinal transection. Thus, it appeared that the brain lesion and the brain itself were not necessary for the production of spinal fixation.

4.5 A SPINAL PHENOMENON?

The new preparation we had begun to investigate seemed to necessitate only the spinal cord and not the brain in the production of spinal fixation. However, there were other questions. Was the retained fixation a product of spinal alterations or of ongoing inputs from the injured tissues around the stimulation sights? Was the alteration in the tissues, the muscles or the central aspects of the reflex circuits in the cord? What were the time parameters of the pure spinal components of the process?

We began to answer these questions in another series of studies (Steinmetz, Beggs, Lupica and Patterson, 1983; Steinmetz, Dobson, Patterson, Cervenka and Romano, 1982; Steinmetz and Patterson, 1985). To look at the involvement of ongoing activity from the cord to the periphery in the retained flexion, we gave 5 groups of rats 10-50 minutes of hind limb stimulation (1-3 ma). We then assessed the asymmetry after stimulus termination, spinalized the animals and gave them a complete dorsal and ventral root rhizotomy from T10-L6. The asymmetry remaining after the rhizotomy was almost identical to that remaining after stimulation had been given to previously rhizotomized animals (see above). Thus, the retained asymmetry was dependent on a continued outflow from the spinal cord. The asymmetry could, however, be due to an alteration of activity at the neuromuscle junction. We gave a group of animals direct stimulation to the severed ventral roots serving the hind limb, this causing a full flexion but without involving the spinal circuits. No appreciable flexion remained in any animals following 50 minutes of ventral root stimulation, despite strong flexion for the full time. It was apparent that the retained flexion or spinal fixation was not a product of alterations at the neuromuscle junction.

With the use of high levels of peripheral stimulation (1-3 ma or more) for 50 minutes, there will be some damage to tissues around the insertion points of the wound clips through which the stimulation is given. We assessed the contribution of continuing inputs from damaged or irritated sensory inputs to the retained flexion in two studies. In each study, animals were anesthetized and prepared for stimulation of the right hind thigh by wound clip application. Laminectomies were performed for subsequent spinal transection and the animals placed in the apparatus as above. In two groups, 20 and 45 minutes of stimulation were given prior to spinal transection and in the other two, after spinal transection. In all groups, following stimulus termination, remaining asymmetry was measured, then lidocaine injected subcutaneously under the wound clips. The dose of lidocaine had been previously shown to be sufficient to preclude any flexion when the stimulus was applied to the wound clips. Thus, the lidocaine dose was sufficient to preclude any sensory input from the area of the wound clips. In both 20-minute groups (transection before and after stimulation) there was only about a gram of retained flexion after lidocaine injection, whereas in both 45-minute groups, there was between 13 and 17 grams of remaining flexion tension. These data clearly showed that the flexion remaining after 45 minutes of stimulation was not due to ongoing sensory input from damaged tissues. In fact, in the 20-minute groups, most animals showed no remaining flexion immediately after the stimulus was terminated. Thus, the necessity of alterations in the spinal circuits seemed assured.

The use of a skin delivered stimulus in the spinal fixation preparation leaves open the possibility that some of the effect could be due to receptor sensitization or other effects in the area of the electrical stimulus delivery. In addition, despite the data cited above that the muscle is not involved in the flexion retention, there could be a contribution of muscle to the final retained asymmetry. To examine these possibilities, we conducted a series of studies (Steinmetz and Patterson, 1985) designed to look at a nerve-muscle and pure nerve preparation of spinal fixation. In the first study, we stimulated the tibial sensory nerve as the sensory input and recorded tension from the severed tendons of the peroneous longus and tibialis cranialis muscles. The nerve was prepared for stimulation by a bipolar wire electrode and bathed in a warmed mineral oil bath. The tendons were attached to a force transducer and output recorded on a polygraph. The animals were warmed and maintained at a surgical level of anesthesia. A spinal transection was performed prior to stimulation. Groups of animals were stimulated for 10, 20,

30 or 40 minutes and the tendon tension recorded. The results showed that the amount of tension remaining after stimulation was in direct proportion to stimulation time. Thus, the 10-minute animals showed almost no retained tension after 5 minutes, whereas the 40-minute animals showed almost no diminution of tension after 5 minutes. The motor nerves were cut after post stimulation measurements and in all groups all remaining muscle tension disappeared.

In the second study of this series, a purely neural preparation was used. Stimulation was delivered to the tibial nerve and responses recorded as summated nerve action potentials from the deep peroneal motor nerve. All animals were spinalized prior to the stimulation. Motor nerve volleys were digitized and recorded by computer for later averaging. The computer sequenced stimulus delivery and recording epochs. Once prepared each animal underwent a series of test and stabilization trials to make sure that the response was stable to a standard stimulus. Once a stable baseline was obtained, animals received either 10, 20, 30, 40 or 50 minutes of sensory nerve stimulation, with a test pulse interspersed during interruptions of the stimulus every minute. Only test pulse nerve volleys were recorded. Following stimulus termination, the test pulse was delivered once a minute for 30 minutes to assess changes from baseline response to the test pulse. Control group animals received the same treatment but no sensory nerve stimulation other than the test pulses. The results (see Figure 4.4) showed that the 10 and 20 minute stimulation groups showed essentially no increased nerve volley amplitude over the controls and were essentially at baseline during the 30 minute retention period. The 30, 40 and 50-minute stimulation groups showed an increased response during retention and in general the response amplitude rose during retention. The 40 and 50-minute groups were at 190% of baseline at the end of the 30-minute retention period. Their controls stayed essentially at baseline. The response latencies of the groups did not show any systematic changes during the study.

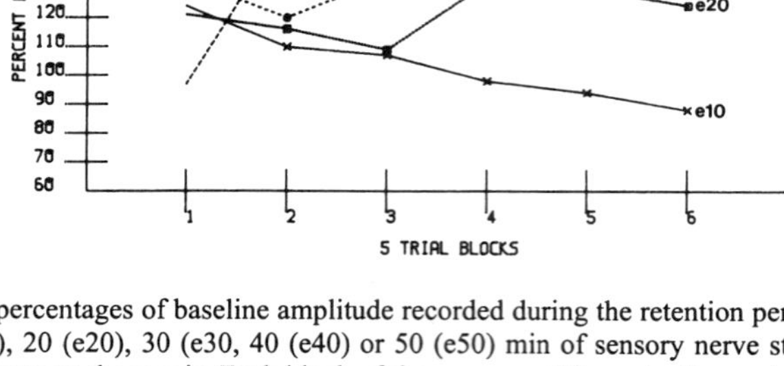

Figure 4.4. Mean percentages of baseline amplitude recorded during the retention periof of experimental rats given 10 (e10), 20 (e20), 30 (e30, 40 (e40) or 50 (e50) min of sensory nerve stimulation (ns = 7). Block number appears on the x-axis. Each block of the average of five retention period trials.) Reprinted from Steinmetz and Patterson, 1985 with permission.

These data showed convincingly that the spinal cord was capable of undergoing a long-lasting increase in excitability in response to sensory input. In addition, the sensory receptors and muscle effectors were not an integral part of the process. Spinal fixation is indeed a phenomenon resident in the spinal cord and not dependent on the brain or on peripheral processes. Spinal fixation can be demonstrated in rats that are given hind limb stimulation prior to spinal transection or after spinal transection. However, differences between animals given spinal transection prior to or after stimulation have been found, as seen in the next section.

4.6 TEMPORAL AND STIMULUS INTENSITY PARAMETERS IN SPINAL FIXATION

As seen above, the time required for fixation of postural asymmetry in rats given peripheral stimulation had been shown to be on the order of 45 minutes. However, we had not examined the issues of temporal effects or of stimulus intensity effects carefully with both spinalized and intact animals.

In the first study to look at stimulus intensity and duration parameters (Steinmetz, Lupica, Secard and Patterson, 1982), we utilized initially two groups of animals given between 1 and 3 ma of peripheral stimulation for 60 or 90 minutes, followed by spinal transection. The results showed that the longer times of stimulation did not produce significantly different asymmetry retention either prior to or following spinal transection. Together with the previous study showing that about 40 minutes of stimulation was necessary to produce consistent retention in a nonspinalized animal, it appeared that with these parameters, 60 minutes of stimulation would produce maximal fixation. However, in the second study of the series, we gave nonspinalized rats 30, 45, 60 and 90 minutes of stimulation at a set intensity of 1, 3 or 5 ma. A measurement of asymmetry retention was taken and the animals spinalized. Following a 5-minute wait, retention was again measured. The results showed that at 30 minutes, almost no retention occurred in the 1 ma group, but some retention was evident in the 3 and 5 ma groups, but it was not consistent across animals. At the longer intervals, all animals in the 3 and 5 ma groups showed retention, but even at 60 and 90 minutes, not all 1-ma animals showed retention following spinal transection.

In another study (Steinmetz, Beggs, Cervenka, Romano and Patterson, 1982) looking at stimulus duration, but in rats spinalized prior to stimulation, rats were given 10, 20, 30, 40, 45 or 50 minutes of 1-3 ma of stimulation. The results showed that consistent retention was found in animals given 30 minutes of stimulation and some animals showed retention at 20 minutes. These results indicated that spinal transection prior to stimulation allowed more rapid fixation of the postural asymmetry. Thus, it is possible that the higher brain centers act to inhibit the formation of spinal fixation. The inhibitory effect could be due to reflex inhibition from higher centers in response to the strong nociceptive inputs from the hind limb stimulation.

In a study reported in our 1982 paper (Steinmetz, et al., 1982), we investigated the possibility that once begun, fixation could consolidate over the SFT without major inputs. We gave intact, anesthetized rats 10, 20, 30, 40 or 50 minutes of hind limb stimulation and allowed each group to rest for the remaining time to 50 minutes. Thus, the 10-minute stimulation group had an additional 40-minute wait prior to asymmetry measurement, spinal transection and remeasurment. In all groups, there was significant retention of asymmetry after spinal transection. This indicated that once begun, the process of fixation could continue without major sensory input. Additional information on this effect was gained in a later study (Steinmetz, et al., 1983) in which we gave three groups of animals 20 minutes of stimulation. One group then had a 30-minute wait and then had lidocaine injected under the wound clips. The second group had the lidocaine injection after the 30-minute wait and the third group had lidocaine injected prior to a 50-minute wait. The group injected prior to the 30-minute wait had little remaining asymmetry, while the group injected after the 30-minute wait had normal fixation for this paradigm. The group injected prior to the 50-minute wait had more asymmetry than the first group but not as much as the second group. These results suggested that the irritation from the stimulation site was contributing to the fixation during the wait period, but the 50-minute injection wait group showed that not all of the effect was due to the site irritation.

In most studies, we measured the retained asymmetry immediately after stimulation (with a 5-minute wait prior to measurement after stimulation or spinal transection). In some cases we have looked at asymmetry retention up to 2 hours after stimulus cessation (e.g., Steinmetz, et al., 1982). To investigate the longevity of the retained asymmetry, we looked at retention after longer periods of time (Steinmetz, Beggs, Molea and Patterson, 1985). In this study, we stimulated intact rats for 45, 60 and 90 minutes, then allowed them to recover in their home cages for 24, 48 or 96 hours. All animals recovered without incident from the stimulation and had normal locomotion in their home cages. After the requisite waiting period, the animals were reanesthetized and a spinal transection performed. All groups and most animals showed a return of asymmetry after the spinal transection. The 45-minute stimulation group had less return of asymmetry than the 60 and 90-minute groups. This study indicated that, as in the Frankstein study (1947), the animals were compensating for the altered spinal excitability with higher centers and once the spinal transection was performed, the spinal effects were evident.

We have pursued the question of stimulation duration and intensity parameters in subsequent unpublished studies, but have consistently found that at a stimulus intensity of .5 ma, little fixation has been noted even with stimulation times of up to 3 hours (Patterson, Anderson, Jenkins, Jensen and Rodman, 1998). With a 1 ma. stimulus intensity, good fixation can be produced by stimulation times of 1.5 –3 hours, but the optimal stimulation intensity appears to be about 2.5 to 3 ma. This intensity produces good retained asymmetry in 30-45 minutes, depending on whether the animals are intact or spinalized prior to stimulation. It is evident from the data that the spinal fixation effect is long lasting, remaining for at least 96 hours, and that it can continue to consolidate after it is initiated. However, at a very low stimulus intensity, even long bouts of stimulation, which produce some flexion, fail to produce significant amounts of fixation. Thus, the effect is a robust one that occurs with expected regularity in response to stimulus duration and intensity parameters.

4.7 OTHER EFFECTS

As noted above, Mouravieff-Lesuisse and his colleagues (1970) looked at the effects of electroconvulsive shock (ECS) on SFT. Steinmetz (1983) also looked at ECS on the fixation process. In this study, we exposed the spinal cord from T7-L2 and transected the cord at T7. All animals received 30 minutes of hind limb stimulation, a 5 minute wait, an asymmetry measurement, a 30 minute no stimulation wait period and another asymmetry assessment. In addition the animals were assigned to a control group, an ECS-0 (ECS immediately before stimulation), ECS 15 (ECS after 15 minutes of stimulation), ECS-30 (ECS immediately after stimulation, ECS 45 (ECS 15 minutes into the wait period) or an ECS-60 (ECS after the wait periods). The ECS was a single 10 ma 500-msec pulse delivered directly to the spinal cord via an insulated insect pin with a flat face applied to the cord surface. The results showed that the animals given ECS at the end of the stimulation period had almost no remaining fixation after the wait period, but that all other groups had significant remaining fixation after the wait period. The results suggested that the fixation process was undergoing a consolidation immediately after stimulus termination, but that by 45 minutes it had become stable. Also, the process was not disrupted by the ECS given midway through the stimulation episode. These results are in keeping with the notion that memory is vulnerable to disruption at an early stage of consolidation, but that once it is stable, it is not vulnerable.

In the late 1980s, another student, Michael Bartelt began in our laboratories, a series of studies looking at other parameters of spinal fixation. In the first of these, he looked at the effects of various stressors on the fixation process (Patterson, Bartelt, Johnson and Howe, 1991; Patterson, Bartelt, Johnson and Jackson, 1992). In several studies, he examined the effects of restraint, swimming and noise stressors on the development of spinal fixation. In all cases, we found that the stress imposed on the animals shortly prior to anesthetization, spinal transection and hind limb stimulation increased the amount of residual postural asymmetry remaining after the fixation period. In these studies, we utilized a 30-minute stimulation period. These studies showed that an increased general level of excitation brought on by the stressors added to the stimulation effects to increase spinal fixation.

Bartelt also investigated the effects of d-amphetamine on the fixation process. It had been found in earlier studies that d-amphetamine seemed to increase fixation. In these studies (Johnson, Bartelt, Johnson and Patterson, 1991; Patterson, Bartelt and Hutchins, 1990; Patterson, Bartelt, Hutchins, Kerns and Guidette, 1990) we generally observed an increased intensity of fixation with the d-amphetamine doses, sometimes to the extent that the animals had up to 3 times the usual fixation. However, an analysis of the findings strongly suggested that the effect was only on the general strength of the effect and not on the actual process, as the time to fixation was not altered with increasing dosage. We concluded that the d-amphetamine was acting only as a general exciter of the system, and had little effect on the actual process. In another study, we looked at the effects of mustard oil application to the foot of the stimulated leg (Patterson, Bartelt, Jackson and Johnson, 1992). As expected the mustard oil application increased fixation presumably by increasing afferent activity to the spinal cord, thus essentially increasing the effective stimulation.

In his doctoral study, Bartelt (1994; Patterson and Bartelt, 1994) investigated the effects of altering hind limb temperature on the fixation process. Alella (1948) had found that cooling guinea pigs with cerebellar lesion induced lesions first increased then decreased and even reversed fixation. Bartelt either cooled or heated the limb following stimulation. Separate groups had the limb temperature maintained at 27, 30, 33.5 (normal) 37 or 40 degrees C. He found that cooling the limb to 30 degrees C increased the fixation amount, but that cooling to 27 degrees C decreased it somewhat from that evidenced at 30 degrees C. Warming the limb decreased fixation from normal but it increased with the 40-degree warming. Core temperature was also significantly altered by the peripheral alterations, and showed increased fixation with either increased to decreased temperatures. The results suggested that the decreased peripheral temperature may have increased c-fiber nociceptive inputs, whereas increasing temperatures may have decreased it up to a point. The studies showed the importance of keeping body and limb temperatures constant during the fixation process.

In a series of studies looking at possible disrupting effects on fixation, Bartelt (1991) investigated the effects of moving the hind limb or interrupting the fixating stimulus during fixation. He found that the act of measuring the fixation amount by stopping the stimulus and stretching the leg to normal position at 20, 25, 30, 35 or 40 minutes into the 40 minute stimulation period (spinalized animals) decreased the amount of final fixation relative to both stimulus interruptions and to the uninterrupted controls. In unpublished observations, we have consistently found that significant movements during the fixation period will decrease fixation amounts. In a puzzling finding (unpublished) we have observed that holding the animal's limb in an extended position during the stimulation period inhibits any flexion following the termination of the stimulus. It is as though the proprioceptive inputs from the limb must signal flexion during the stimulation or no flexion will be fixated.

4.8 NEURAL INTERACTIONS

In the early 1990s we began a series of studies looking at various neural interactions that could effect the fixation process. Initially, we hypothesized that spinal fixation at one locus (right hind limb area of the spinal cord) should be affected by inputs from other areas. Thus, is seemed logical that if a short stimulus was given to the hind limb, insufficient to produce frank postural asymmetry, there should be still some amount of neural excitability enhancement occurring. A second bout of stimulus input at a later time should then produce more fixated asymmetry than expected by the second stimulation session alone. That is, the two stimulation periods should be additive. In a series of studies, we have found this not to be the case (Patterson and Knewtson, 1996; Patterson, Mullauer, Suddarth, Sipe and House, 1997; Patterson, Perkins, Feizelli and Bartelt, 1993; Patterson, Woody and Patterson, 1995). In these studies, we first stimulated, in intact animals, the right hind limb, then following some delay, stimulated the limb again. Persisting asymmetry in the doubly stimulated condition was assessed against the persisting asymmetry in groups treated the same except that no stimulation was given during the first period. Results of these studies have been consistent in showing that when the animals received stimulation during the first period, the resulting retained flexion after the second stimulation period is reduced relative to that in subjects only receiving stimulation during the second period. The one exception to this seems to be when the second stimulation intensity is greater than that during the first stimulation bout. We have tested this effect with 1.5 and 3.5 ma of stimulus intensity and for periods between stimulus bouts of from 1 to 9 days. It appears that if the initial stimulus is less intense than the second stimulus the blocking will last for about 6 days, whereas if the initial stimulus is more intense, the blocking effect can last for 7 to 8 days. The effect was not observed after 9 days separated the two stimulus periods. This blocking effect is interesting in that it is opposite the expected effect. Rather than adding together, the initial stimulus bout seems to interfere with the fixation of asymmetry by the second stimulus bout. In an elegant study, Mantyh and his associates (Mantyh, et al., 1995) have reported that short bouts of stimulation can cause removal of receptor sites from second order interneuron dendrites in the spinal cord. It is possible that the initial bout of stimulation causes some comparable process in the fixation preparation that inhibits the establishment of fixation to the second stimulus. It seems possible that this could be an adaptive process to limit the neural alterations in spinal networks produced by successive stimulus events.

In recent studies (Dally and Patterson, 2000; Patterson, Hallahan and Anderson, 1999; Thompson, Patterson and Ullman, 2000) we have begun to investigate the effects of stimulating the opposite front and hind limbs on fixation of the right hind limb. It is possible that activation of the opposite hind limb could, through the crossed extensor reflex, inhibit the flexion fixation of the right limb. In contrast, stimulation of the opposite fore limb could enhance the fixation of the right hind limb through opposite crossed flexor reflex activation. In the first study, we found that fixation of the right hind limb can be reduced a small amount when followed by stimulation of the left hind limb. In addition, concurrent stimulation of the left and right hind limbs produces a small amount of reduction in the expected flexion fixation of the right limb, but concurrent stimulation of the opposite fore limb was found to have no significant enhancing effect on the right hind limb flexion retention.

4.9 OTHER STUDIES

In a series of studies looking at the underpinnings of spinal fixation (Anderson, Mokler and Winterson, 1991; Anderson and Winterson, 1995; Moore, Mokler and Winerson, 1992), Winterson and Anderson have examined the role of various substances in spinal fixation. They have focused on the role of N-methyl-D-aspartate (NMDA) in the fixation process and have found that intrathecally applied NMDA produced more fixation, whereas MK-801 (an NMDA inhibitor) reduced it. In addition, they found that the MK-801 administered 72 hours following stimulation had no effect. Capsaisan administration to reduce nociceptor input, decreased fixation in both neonates and adult animals when measured immediately after stimulation, but in adults, a normal amount of fixation remained following 72 hours. They also found that 5-hydroxytryptamine (5-HT) seemed to be involved in suppression of fixation in intact animals, presumably by descending serotonergic fibers which exert an effect through 5-HT2 receptors. They have hypothesized that NMDA receptors are activated during stimulation and are part of the cascade of events that lead to the retention of asymmetry. Their studies have also shown that fixation can last for up to several days.

Russian Studies of Spinal Fixation

Russian efforts in working with spinal fixation began about the same time, as did our work, in the late-1970s. Professor Varatanian and his group worked in this area but along a somewhat different line, and neither group knew about the work of the other until fairly recently. The Russian investigations were directed more toward the molecular basis of this phenomenon in models with central ablations used to produce the fixation of the central excitability. These investigations showed that the production of centrally induced fixation apparently resulted in the appearance of neurohumoral factors that could be extracted from brain tissue and when injected intraperitonealy into newly lesioned animals, would shorten the spinal fixation time (SFT) (Bogdanov, Mikhailenok and Avaliani, 1987). The details of the extraction technique and preliminary studies on enzymatic degradation led the authors to the supposition of an oligopeptidic nature of these factors. In later investigations in the Pavlovian Physiology Department of the Institute for Experimental Medicine of the Academy of Medical Sciences of the USSR, it was determined that these factors seem able to induce unilateral hind limb flexion (PA) in spinal animal-recipients (Vartanian, Moroz and Silakov, 1980; Vartanian, Varlinskaya, Tsikunov and Shkliaruk, 1982; Vartanian and Yu.V., 1978) when injected into the lumbar segments of spinal cord. These substances, named postural asymmetry factors (PAF), were confirmed to be oligopeptides (Vartanian and Yu.V., 1978) with molecular weight lower than 2 kDa, containing lys- or arg- (since it's activity was sensitive to trypsine). Several fractions of these peptides were discovered, one of them being identified as arginine-vasopressin (Vartanian, et al., 1982). It was playing the leading role in the right spinal cord hemisection model. PAF activity was identified in cerebrospinal fluid (CSF) of patients with unilateral motion disorders resulting from stroke, brain trauma or tumor (Makarov, et al., 1991;

Vartanian, et al., 1982). More recently, the Russian workers have identified several PAF fractions (as many as 10) which are usually balanced in the normal individual. With damage to one side of the brain or cord, the PAF fractions change, with some becoming much less predominant than others (Klementyev, et al., 1986). It is possible that a greater understanding of the role of PAFs in brain function and balance may lead to methods to increase rehabilitation in CNS injured patients. Other studies of this group have confirmed the results of some of Patterson's work. In additional studies, the Russian group has observed that fixation produced by several methods on one side of the animal (for example, on the right leg) shifts to the opposite side after about 3 hours (Freidenfeld, Danilovskii, Ptitsyna and Golikov, 1990). This is an effect not observed in the American work to this time, and is a very important observation that could be caused by shifts in the PAF balances during the intervals following the initiating central lesions. It is possible that the shift of the observed asymmetry from one side to the other could reflect some alteration in body temperature or condition that has not occurred in our studies.

Since the contacts with the Russian group were established in the mid 1990s, with the death of Professor Varatanian, we have not had additional contact with the group in St. Petersburg.

4.10 SUMMARY

The spinal fixation phenomenon has been shown to be a property of the spinal cord. It is produced by stimulation of the peripheral limb or by direct stimulation of the sensory nerve, as well as by central lesions of various types. It has been shown to be a long lasting alteration in spinal neural excitability strong enough to support active limb flexion in the rat. However, in a pilot study with a limited number of cats, while we were able to see increased neural excitability in a neural preparation, we could not elicit frank hind limb flexion (unpublished data). It follows generally regular patterns of increasing strength with increased stimulus intensity and duration. It appears to be affected by descending influences, probably mainly inhibitory, from the higher brain centers and small effects on fixation can be found from other spinal inputs. Data available suggest that NMDA receptors may play a role in its formation. The Russian data suggest that there may be substances elaborated during fixation formation that can influence it course, and affect the course of fixation in recipient animals.

There is now an increasing amount of data available on the cellular basis of spinal plasticity in various preparations. Much of these data come from studies of inflammatory models of spinal excitability changes. For a recent review of this area, see the recent National Academy of Sciences Symposium (Dubner and Gold, 1999). It appears that there are basically four stages of spinal excitability alterations: 1. Short-term – sensitization; 2. Intermediate term – long-term sensitization; 3. Semipermanant – fixation and alterations produced by peripheral inflammation; and 4. Relatively permanent – inhibitory cell death and perhaps sprouting of new excitatory synapses. The spinal fixation model fits into the third of these stages and is likely underlain by many of the processes now being elucidated for nociceptor produced changes in the spinal cord.

The spinal fixation phenomenon is now well characterized and is easily produced in the rat model. It is a paradigm that is not confounded by changes in peripheral receptors and can be measured either by neural responses or muscle contraction. It thus appears to be an ideal candidate for further studies of spinal plasticity and its neural basis.

4.11 ACKNOWLEDGEMENTS

The research reviewed here was supported by the National Institute of Neurological Disorders and Stroke, the American Osteopathic Association Bureau of Research, the Ohio University College of Osteopathic Medicine and the University of Health Sciences College of Osteopathic Medicine. We appreciate the help of Brenda Riepenhoff in preparation of the manuscript. The author may be reached at drmike@nova.edu

4.11 REFERENCES

Alella, A. Persistence of asymmetries of cerebral origin in the extremities of animals with cord section. *Archivo de Fisiologia,* 1948; 47: 105-112.

Anderson, M.F., Mokler, D.J., and Winterson, B.J. Inhibition of chroninc hindlimb flexion in rat: evidence for mediation by 5-hydroxytryptamine. *Brain Research,* 1991; 541:216-224.

Anderson, M.F., and Winterson, B.J. Properties of peripherally induced persistent hindlimb flexion in rat: involvement of N-methyl-D-aspartate receptors and capsaicin-sensitive afferents. *Brain Research*, 1995; 678:140-150.

Bartelt, M.J. *Effects of repeated measures and stimulus interruptions on spinal fixation in rats.* Unpublished Masters Thesis, 1991.

Bartelt, M.J. *Effects of altering hind leg temperature on spinal fixation in rats.* Unpublished Doctoral Dissertation, 1994.

Bogdanov, O.V., Mikhailenok, E.L., and Avaliani, T.V. Afferent determination of transfer factor formation in the rat spinal cord. *Bulletin of Experimental Biology and Medicine, 1987;104*: 1340-1343.

Carew, T.J. Pinsker, H.M., and Kandel, E.R. Long-term habituation of a defensive withdrawal reflex in aplysia. *Science.* 1972; 175:451-454.

Chamberlain, T.J. *Fixation of experience in the rat spinal cord.* Unpublished Doctoral Dissertation, 1961.

Chamberlain, T.J. *Fixation of experience in the rat spinal cord and its suggested dependence upon RNA.* Unpublished Masters Thesis, 1963.

Chamberlain, T.J., Halick, P., and Gerard, R.W. Fixation of experience in the rat spinal cord. *Journal of Neurophysiology*, 1963; 26:662-673.

Chamberlain, T.J. Rothschild, G.H., and Gerard, R.W. Drugs affecting RNA and learning. *Proceedings of the National Academy,* 1963; 49.

Daliers, J., and Giurgea, C. Effect of brain extracts on the fixation of experience in the rat spinal cord. *Akademiai Kiado, Budapest*, 1969; 191-197.

Dally, L., and Patterson, M.M. Effects of crossed extensor activation on spinal fixation in rats. *Journal of the American Osteopathic Association,* 2000; 100(9):584.

Denslow, J. An analysis of the variability of spinal reflex thresholds. *Journal of Neurophysiology,* 1944; 7:207-215.

Denslow, J.S., Korr, I.M., and Krems, A.D. Quantitative studies of chronic facilitation in human motoneuron pools. *American Journal of Physiol*ogy, 1947; 105 (*2)*:229-238.

DiGiorgio, A.M. Persistenza nell'animale spinale, di asimmetrie osturali e motorie di origine cerebellare. *Archivo de Fisiologia,* 1929 ;27:518-580.

DiGiorgio, A.M. Richerche sulla persistenza dei fenomini cerebellari nell'animale spinale. *Archivo de Fisiologia,* 1943; 43:47-63.

DiGiorgio, A.M. Asimmetria funzionale del midollo spinale da innervazione asimmetrica cerebrale o cerebrale o cerebellare (richerche in animale curarizzatie). *Bollotin Society de Italian Biology Sper.*, 1947a; 23:287-288.

DiGiorgio, A.M. Persistenza di fatti asimmetrici (da lesioni cerebellari o cerebrali) dopo il taglio del midollo in arti previamente deafferentati. *Bollotin Society de Italian Biology Sper.*, 1947b; 23:287-288.

DiGiorgio, A.M. Sui rapporti cerebello-spinali. Asimmetrie cerebellari dopo deafferentazione e sezione del midollo spinale. *Archives Neerland Physiology,* 1947c; 28:466-474.

Dubner, R., and Gold, M. The neurobiology of pain. *Proceedings of the National Academy of Sciences,* 1999; 96(July 6):7627-7630.

Frankstein, S.I. One unconsidered form of the part played by the nervous system in the development of disease. *Science,* 1947;106:242.

Freidenfeld, D.A., Danilovskii, M. ., Ptitsyna, I.B., and Golikov, Y.P. Unilateral peripheral influences on the motor system as activators of postural asymmetry factors in animal with an intact nervous system. *Bulletin of Experimental Biology and Medicine,* 1990; 110:1472-1474.

Giulio, L. Sulla funzione del midollo spinale. Persistenza di saimmetrie da exxitamento labirintico. *Bollotin Society de Italian Biology Sper.,* 1952; 28:1651-1652.

Giurgea, C. The "Nootropic" approach to the pharmacology of the integrative activity of the brain. *Conditional Reflex,* 1973; 8 (2):108-115.

Giurgea, C., Daliers, J., and Rigaux, M.L. Pharmacological studies on an elementary model of learning - the fixation of an experience at spinal level: Part II: Specific shortening of the spinal cord fixation time (SFT) by a brain extract. *Archives International, de Pharmacodynamie et de Therapie,* 1971; 191 (2): 292-300.

Giurgea, C., and Mouravieff-Lesuisse, F. Pharmacological studies on an elementary model of learning - the fixation of an experience at spinal level: Part I: Pharmacological reactivity of the spinal cord fixation time. *Archives International Pharmacodynamics, de Pharmacodynamie et de Therapie* 1971; 191 (2): 279-291.

Giurgea, C., Mouravieff-Lesuisse, F., and Daliers, J. Spinal fixation time as a model for central plasticity. *Paper presented at the IV International Neurobiological Symposium, Magdebourg, May, 1973.*

Jacobson, A.L., Fried, C., and Horowitz, S.D. Planarians and memory. *Nature*, 1966; 209(5023):599-601.

Johnson, E.S., Bartelt, M.J., Johnson, E.D., and Patterson, M.M. Dose-response effects of d-Amphetamine on persisting hind limb flexion (spinal fixation) in rats. *Society for Neuroscience Abstracts,* 1991; 416.12:1046.

Klementyev, B.I., Molokoedov, A.S., Bushuev, V.N., Danilovskii, M.A., Sepetov, .F., Tokarev, A.V., Titov, M.I., and Vartanian, G.A. Purification of the posture asymmetry factor at right side hemisection of spinal cord. *Doklady Akademii Nauk SSSR (in Russian).,* 1986; 291:737-741.

Korr, I.M. (1947). The neural basis of the osteopathic lesion. *Journal of the American Osteopathic Association*, 191-198.

Makarov, A.Y., Vartanian, G.A., Pomnikov, V.G., Chernaev, S.G., Tokarev, A.V., and Shatik, S.V. Neurochemical factors of central motional deseases pathogenesis. *The Journal of Neuropathology and Psychiatry. (In Russian),* 1991; 10:3-6.

Manni, E. Localizzanzioni cerebellari corticali nella cavia. Nota 1: Il "corpus cerebelli". *Archivo de Fisiology,* 1950; 49:213-237.

Mantyh, P.W., DeMaster, E., Malhotra, A., Ghilardi, J.R., Rogers, S.D., Mantyh, C.D., Liu, H., Basbaum, A.I., Vigna, R., Maggio, J.E., and Simone, D.A. Receptor endocytosis and dendrite reshaping in spinal neurons after somatosensory stimulation. *Science,* 1995; 268(16 June): 1629-1632.

Moore, R.D., Mokler, D.J., and Winterson, B.J. Intrathecally administered N-methy-D-aspartate increases persitent hindlimb flexion in rat. *Neuroscience Letters*, 1992; 146(2):223-226.

Mouravieff-Lesuisse, F., and Giurcea, C. Influence of elctro-convulsive shock on the fixation of an experience at spinal level. *Archives Internationales De Pharmacodynamie et de Therapie*, 1970; 183(2):410-411.

Mouravieff-Lesuisse, F., and Giurgea, C.E. Cortical regulation of the fixation of an experience at the spinal Level. *Unpublished*, 1973:1-11.

Palmer, G., Ward, J.W., and Davenport, G.R. Biochemical and pharmacological investigation of abnormal spinal reflex activity. *Anatomical Record,* 1968; 160:405.

Palmer, G.C., and Davenport, G.R. Involvement of phospholipids in the fixation of abnormal spinal reflexes in newborn rats. *Brain Research,* 1969; 13:394-396.

Palmer, G.C., Davenport, G.R., and Ward, J.W. The effect of neurohumoral drugs on the fixation of spinal reflexes and the incorporation of uridine into the spinal cord. *Psychoplarmacologia, 1970; 17*:59-69.

Palmer, G.C., Davenport, G.R., and Ward, J.W. Involvement of RNA synthesis in the fixation of abnormal reflexes in the newborn rat. *Experimental Neurology,* 1970; 26:263-274.

Palmer, G.C., Davenport, G.R., and Ward, J.W. Involvement of protein synthesis in the fixation of spinal reflexes in the newborn rat. *Brain Research,* 1970; 17:372-375.

Patterson, M.M. Mechanisms of classical conditioning and fixation in spinal mammals. *Advances in Psychobiology,* 1976; 3:381-436.

Patterson, M.M. Classical conditioning of spinal reflexes. In R.F. Thompson, Hicks, H.H., Shvyrkov, V.B. (Eds), *Neural Mechanisms of Goal-Directed Behavior and Learning* (pp. 263-272). New York: Academic Press, 1980.

Patterson, M.M., Anderson, K., Jenkins, H., Jensen, D., and Rodman, D. Effects of long stimulation times and opposite limb stimulaiton on spinal fixaito in rats. *Journal of the American Osteopathic Association,* 1998; 98(7):392.

Patterson, M.M., and Bartelt, M.J. Effects of hindlimb heating and cooling on spinal fixation in rats. *Society for Neuroscience Abstracts,* 1994; 158.10:124.

Patterson, M.M., Bartelt, M.J., and Hutchins, A.G. Effect of d-Amphetamine on spinal fixation in rats. *Journal of the American Osteopathic Association*, 1990; 90(8):723.

Patterson, M.M., Bartelt, M.J., Hutchins, A.G., Kerns, L.E., and Guidette, J.M. D-Amphetamine increases strength, but does not decrease latency of spinal fixation in rats. *Journal of the American Osteopathic Association,* 1990; 90(10):936.

Patterson, M.M., Bartelt, M.J., Jackson, T., and Johnson, E.S. Mustard oil application increases persisting hind limb flexion (spinal fixation) in rats. *Society for Neuroscience Abstracts,* 1992; 363.11: 871.

Patterson, M.M., Bartelt, M.J., Johnson, E.S., and Howe, A. Restraint stress increases induced persisting hindlimb flexion (spinal fixation) in rats. *Journal of the American Osteopathic Association,* 1991; 91(9):904.

Patterson, M.M., Bartelt, M.J., Johnson, R.S., and Jackson, T. Environmental stressors increase persisting hind leg flexion (spinal fixation) in rats. *Journal of the American Osteopathic Association,* 1992; 92(10):1287.

Patterson, M.M., Hallahan, T., and Anderson, K. A weak effect of crossed-extensor reflex on reversing spinal fixation in rats. *Journal of the American Osteopathic Association,* 1999; 99(8):423.

Patterson, M.M., and Knewtson, T. Temporal effects of prior stimulation on later blocking of spinal fixation in rats. *Journal of the American Osteopathic Association*, 1996; 96(9):552.

Patterson, M.M., Mullauer, T., Suddarth, J., Sipe, B., and House, M. Limits to temporal effects of prior stimulaiton on later blocking of spinal fixation in rats. *Journal of the American Osteopathic Association,* 1997; 97(9):548.

Patterson, M.M., Perkins, K., Feizelli, M., and Bartelt, M.J. Inhibition of persisting hat hindlimb flexion (spinal fixation) by previous hind leg stimulation. *Journal of the American Osteopathic Association,* 1993; 93(9):948.

Patterson, M.M., Woody, S., and Patterson, M.S. Effects of prior stimulation on later spinal fixation in rats. *Society for Neuroscience Abstracts,* 1995; 158.10:124.

Sherrington, C.S. *The integrative action of the nervous system.*: Yale University Press, 1905.

Shurrager, P.S., and Culler, E. Conditioning in the spinal dog. *Journal of Experimental Psychology,* 1940; 26:133-159.

Steinmetz, J.E. *A model system approach for the study of memory processes: Fixation of spinal reflex alterations in spinalized rats.* Unpublished Doctoral Dissertation, 1983.

Steinmetz, J.E., Beggs, A.L., Cervenka, J., Romano, A.G., and Patterson, M.M. Temporal parameters involved in retention of reflex alterations in spinal rats. *Journal of Comparative and Physiological Psychology,* 1982; 96(2):325-327.

Steinmetz, J.E., Beggs, A.L., Lupica, C.R., and Patterson, M.M. Effects of local anesthesia on persistence of peripherally induced postural asymmetries in rats. *Behavioral Neuroscience,* 1983; 97(6): 921-927.

Steinmetz, J.E., Beggs, A.L., Molea, D., and Patterson, M. Long-term retention of a peripherally induced flexor reflex alteration in rats., 1985; 327:312-315.

Steinmetz, J.E., Cervenka, J., Robinson, C., Romano, A.G., and Patterson, M.M. Fixation of spinal reflexes in rats by central and peripheral sensory input. *Journal of Comparative and Physiological Psychology,* 1981; 95(4):548-555.

Steinmetz, J.E., Dobson, J., Patterson, M.M., Cervenka, J., and Romano, A.G. Central and peripheral influences on retention of postural asymmetry in rats. *Journal of Comparative and Physiological Psychology,* 1982; 96(1):4-11.

Steinmetz, J.E., Lupica, C.R., Secard, C.M., and Patterson, M.M. Effects of manipulating stimulation intensity and duration on fixation of a peripherally-induced spinal reflex alteration in rats. *Physiology and Behavior*, 1982; 29(6):1039-1044.

Steinmetz, J.E., and Patterson, M.M. Fixation of spinal reflex alterations in spinal rats by sensory nerve stimulation. *Behavioral Neuroscience,* 1985; 99(1):97-108.

Thompson, R.F., and Spencer, W.A. Habituation: A model phenomenon for the study of neuronal substrates of behavior. *Psychological Review*, 1966; 173(1):16-43.

Thompson, S., Patterson, M.M., and Ullman, A. Effects of opposite forelimb stimulation on hindlimb spinal fixation in rats. *Journal of the American Osteopathic Association,* 2000; 100(9):584.

Vartanian, G.A., Moroz, B.T., and Silakov, V.L. Brain cortex model of posture asymmetry and the possibility to manage of stable pathological state. In: The theoretical basis of pathological states. *Leningrad: Medicine, (in Russian).*, 1980; 48-51.

Vartanian, G.A., Varlinskaya, E.I., Tsikunov, S.G., and Shkliaruk, S.P. Neurophysiological and neurochemical mechanisms of functions and states self-regulation. In: Self-regulation of functions and states. *Leningrad: Nauka, (in Russian)*, 1982:37-42.

Vartanian, G.A., and Yu.V.B. Induction of postural asymmetry in an intact recipient by brain extract from a donor with the same syndrome. *Bulletin of Experimental Biology and Medicine, 1978; 86*:996-999.

5 SPINAL CORD PLASTICITY IN THE ACQUISITION OF A SIMPLE MOTOR SKILL

Jonathan R. Wolpaw
Wadsworth Center
New York State Department of Health
and
State University of New York
Albany, New York 12201

5.1 ABSTRACT

Studies of the plasticity underlying the acquisition of motor skills traditionally focus on supraspinal areas such as cerebral cortex and cerebellum. However, a large body of clinical and laboratory evidence indicates that substantial plasticity occurs in the spinal cord and that it plays an important part in skill acquisition. Operant conditioning of the spinal stretch reflex (SSR) or of its electrical analog, the H-reflex, is a simple example of such plasticity and provides a convenient laboratory model for defining the responsible mechanisms. The SSR and the H-reflex are mediated by a wholly spinal and largely monosynaptic pathway consisting of the Ia afferent fiber from the muscle spindle, its synapse on the spinal motoneuron, and the motoneuron itself. Monkeys, humans, and rats can gradually increase or decrease SSR or H-reflex amplitude when rewarded for doing so. Driven by task-appropriate change in descending influence from supraspinal structures, reflex change begins with a rapid change (phase I) in the first 1-2 days and then continues with gradual change (phase II) over weeks and months. Reflex change is accompanied by plasticity at multiple sites in the spinal cord. Plasticity occurs in motoneuron firing threshold and conduction velocity, in several different synaptic terminal populations on the motoneuron, and probably in interneurons as well. It also occurs contralaterally. The corticospinal tract is probably essential for producing this plasticity. While some changes, such as that in firing threshold, probably contribute to the rewarded behavior (primary plasticity), others may preserve previously learned behaviors (compensatory plasticity), or are activity-driven products of plasticity elsewhere (reactive plasticity). As these data and data from other simple vertebrate and invertebrate models illustrate, a complex pattern of plasticity is the necessary and inevitable outcome of even the simplest learning. Operant conditioning of the SSR and H-reflex appears to be similar to changes in spinal cord function that occur in early life and in the course of skill acquisition throughout life. The gradual development of this plasticity over the course of training is an important element in skill acquisition. The spinal cord is shaped by long-term descending influence so as to provide a substrate suited for execution of an individual's entire roster of motor performances.

5.2 INTRODUCTION

Recent research indicates that the capacity for activity-driven plasticity is ubiquitous in the central nervous system, occurring at sites from neocortex to spinal cord (Maren and Baudry, 1995; Calabresi, Pisani, Mercuri and Bernardi, 1996; Oda, Charpier, Murayama, Suma and Korn, 1995; Muir and Steeves, 1997; Fisher, Fischer and Carew, 1997). In spite of this new knowledge, investigations of the plasticity underlying the acquisition of motor skills continue to focus almost exclusively on supraspinal areas, particularly cerebellum and sensorimotor cortex. Despite extensive clinical and laboratory evidence that modifications in descending input can produce lasting changes in spinal cord neurons and pathways ((Kandel and Spencer, 1968; Mendell, 1984; Wolpaw and Tennissen, 2001) and this volume for review), the importance and role of spinal cord plasticity in skill acquisition remains largely unrecognized and unexplored. This is unfortunate first because it hampers efforts to understand the processes of skill acquisition and second because it ignores a potentially valuable model system for studying CNS plasticity in general. Because the spinal cord is more technically accessible than the rest of the CNS, and because its inputs and outputs are well-defined and largely controllable, the spinal cord presents an opportunity to localize and describe the mechanisms of plasticity that is almost unique in vertebrates and rivals the opportunities provided by invertebrate models.

The spinal stretch reflex (SSR), or tendon jerk, is the simplest behavior of the vertebrate CNS. As illustrated in Figure 5.1, it is mediated mainly by a two-neuron, monosynaptic pathway composed of the Ia afferent fiber, its synapse on the alpha motoneuron, and the alpha motoneuron (Magladery, Porter, Park and Teasdall, 1951; Matthews, 1972; Henneman and Mendell, 1981). The H-reflex, also shown in Figure 1, is the electrical analog of the SSR (Figure 5.1). In the H-reflex, the Ia afferent fiber is excited by direct nerve stimulation rather than by muscle stretch. The wholly spinal pathway of the SSR and H-reflex is influenced by descending activity from supraspinal structures, and this influence can change reflex amplitude. Studies over the past two decades have shown that, when exposed to a protocol in which reward depends on SSR or H-reflex amplitude, monkeys, humans, and rats can modify this descending influence so as to gradually increase or decrease the reflex (Figure 5.2) (Wolpaw, Braitman and Seegal, 1983; Wolpaw, 1987; Evatt, Wolf and Segal, 1989; Chen and Wolpaw, 1995; Wolf, Segal, Heter and Catlin, 1995). The gradual acquisition of this simple skill, i.e., a larger or smaller SSR or H-reflex, involves plasticity in the spinal cord itself, since evidence of it remains even if all descending influence is removed (Wolpaw and Lee, 1989). Thus, this phenomenon, referred to as operant conditioning of the SSR or H-reflex, provides an opportunity to define the plasticity responsible for a simple motor skill, the processes that create and maintain this plasticity, and the manner in which this plasticity translates into behavior. This chapter reviews the phenomenon of SSR or H-reflex conditioning, summarizes current knowledge of the spinal cord plasticity that accompanies it and the descending influence that creates it, and discusses its relationships to the acquisition of more complex motor skills.

5.3 SPINAL STRETCH REFLEX (SSR) AND H-REFLEX CONDITIONING

The SSR and the H-reflex can both be operantly conditioned, and their conditioning is comparable in rate, final magnitude, and other major features (Wolpaw, Braitman, and Seegal, 1983; Wolpaw, 1987; Evatt, Wolf, and Segal, 1989; Chen and Wolpaw, 1995; Wolf, Segal, Heter and Catlin, 1995). SSR conditioning in humans may furnish a new method for reducing the spinal reflex abnormalities that accompany numerous chronic CNS disorders (Segal and Wolf, 1994; Wolf and Segal, 1990; Segal, 1997). At the same time, H-reflex conditioning is a more practical laboratory model, because the H-reflex bypasses the muscle spindle and is readily incorporated into long-term recording in unrestrained animals. Originally reported in monkeys, H-reflex conditioning is now studied mainly in rats.

Figure 5.2A shows the protocol for operant conditioning of the H-reflex in rats. The primate H-reflex and SSR protocols are comparable to the rat protocol in all important aspects. A key feature of the protocol is that the reflex is elicited at an unpredictable time. Because of this, and because the H-reflex is the earliest possible CNS response, the animal can change H-reflex amplitude only by maintaining continual appropriate descending influence over the spinal arc of the reflex. By linking reward to H-reflex amplitude, the protocol operantly conditions the animal to maintain this influence. The presence of this descending influence through the period of conditioning gradually changes spinal cord function and thereby increases (i.e., HR↑ conditioning mode) or decreases (i.e., HR↓ conditioning mode) the H-reflex. As illustrated in Figure 5.2B, the rat soleus H-reflex increases markedly under the HR↑ mode and decreases markedly under the HR↓ mode. As Figure 5.2C illustrates, conditioning in monkeys and humans produces similar results (Wolpaw, Braitman and Seegal, 1983; Wolpaw and Herchenroder, 1990; Chen and Wolpaw, 1995; Wolf, Segal, Heter and Catlin, 1995).

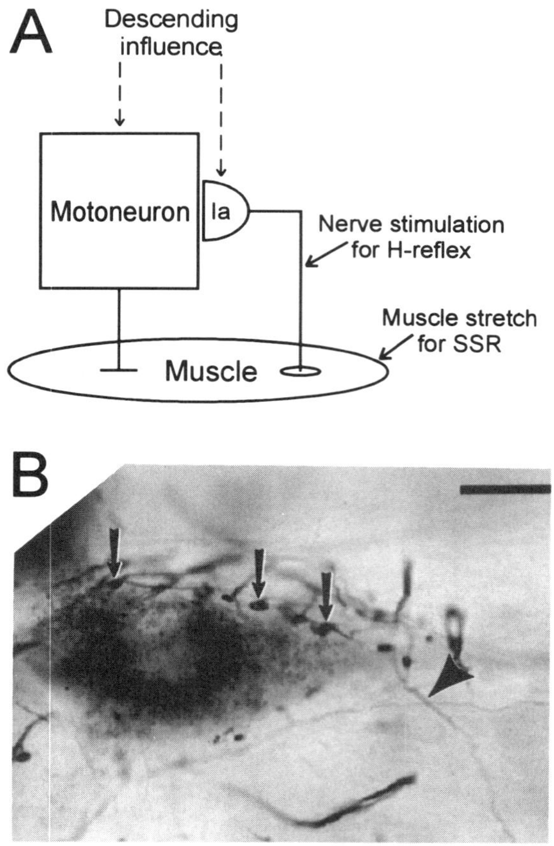

Figure 5.1. The pathway of the spinal stretch reflex (SSR) and its electrical analog, the H-reflex. (A) The pathway consists of the Ia afferent neuron from the muscle spindle, its synapse on the alpha motoneuron, and the motoneuron itself. When the Ia afferent is excited, it excites the spinal motoneurons that innervate the same muscle and its synergists. If the afferent is excited in the normal way, i.e., by muscle stretch, the muscle's response is the SSR. If it is excited by direct electrical stimulation of the nerve, the response is the H-reflex (Brown, 1984). The SSR and the H-reflex are assessed by EMG or by their mechanical effects. While their pathway is completely spinal, it is influenced by activity in descending pathways from supraspinal regions. This activity can affect the Ia terminal (presynaptically) and/or the motoneuron. The SSR is also affected by descending control of muscle spindle sensitivity (Matthews, 1972; Baldissera, Hultborn and Illert, 1981). (B) Optical section showing a rat soleus motoneuron labeled with cholera toxin - horseradish peroxidase injected into the muscle, a primary (probably Ia) afferent fiber (arrowhead) labeled with biocytin injected into the dorsal root, and several putative synaptic terminals (arrows) of that fiber on the motoneuron. The bar is 10 μm. (Micrograph from Feng-Chen, Chen and Wolpaw, 1998.)

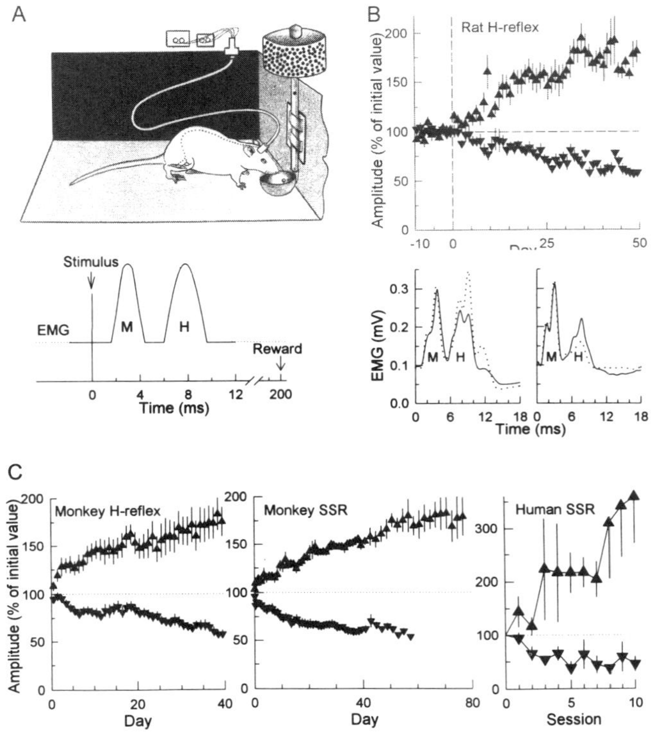

Figure 5.2. The conditioning protocol and its effects. (A) Soleus EMG is measured continuously in a rat with chronically implanted EMG electrodes and a tibial nerve cuff. Whenever EMG absolute value stays in a specified range for a randomly varying 2.3-2.7 s period, a nerve cuff stimulus produces a threshold M response (i.e., a direct muscle response (Brown, 1984)) and an H-reflex. For the initial 10 days, the rat is exposed to the control mode, in which no reward is given. The H-reflex is simply measured to define its initial amplitude. For the subsequent 50 days, the rat is exposed to the HR↑ or HR↓ conditioning mode, in which a food-pellet reward is given when the H-reflex is above (HR↑mode) or below (HR↓ mode) a criterion value. Background EMG and M response stay constant throughout. Successful conditioning, which is defined as a change of at least 20% in the correct direction (Wolpaw, Herchenroder and Carp, 1993; Chen and Wolpaw, 1995), occurs in more than 75% of the animals. Their data are displayed in (B). In the top graph are average daily H-reflexes (±SE) from 9 HR↑ rats (▲) and 12 HR↓ rats (▼) for the control-mode exposure (i.e., days -10 to 0) and for the subsequent HR↑ or HR↓-mode exposure (i.e., days 0 to 50). Under the HR↑ mode, the H-reflex increases gradually to about 175% of its initial value, while under the HR↓ mode it decreases gradually to about 60%. The bottom graphs display average poststimulus EMG for representative days from an HR↑ rat (left) and an HR↓ rat (right) under the control mode (solid) and near the end of HR↑ or HR↓ exposure (dashed). The H-reflex is considerably larger after HR↑ conditioning and considerably smaller after HR↓ conditioning. Background EMG (shown here by EMG at zero time) and M responses do not change. When a rat is switched from the HR↑ to the HR↓ mode (or vice versa), the H-reflex change reverses in the same gradual way (Chen and Wolpaw, 1996). (C) Average results for up conditioning (▲) and down conditioning (▼) of the triceps surae H-reflex in monkeys (left), the biceps brachii SSR in monkeys (middle), and the biceps brachii SSR in humans (right). The courses and final magnitudes of change are comparable to those described in the rat. B from Chen and Wolpaw (1995, 1996) and unpublished data, C from Wolpaw, Maniccia and Elia (1994), Wolpaw and O'Keefe (1984), and Wolf, Segal, Heter, and Catlin (1995).

5.4 THE TWO-PHASE COURSE OF OPERANTLY CONDITIONED REFLEX CHANGE

Average curves for the development of reflex change were calculated, first for SSR conditioning in monkey biceps and subsequently for H-reflex conditioning in monkey triceps surae and rat soleus. These curves showed that conditioned change developed in two phases, a rapid phase I that occurred in the first 1-2 days and a slow phase II that continued for weeks (Wolpaw and O'Keefe, 1984; Wolpaw, Moniccia and Elia, 1994, Wolpaw, Chen, Chen and Wolpaw, 1999). As illustrated in Figure 3, phase I produces an abrupt change in reflex amplitude (e.g., for the monkey SSR, the phase I rate of change is more than 10 times the phase II rate). It presumably results from rapid task-appropriate change in descending influence over the spinal arc of the reflex. The nature of this change was illuminated by the observation that day-to-day variation in SSR amplitude falls abruptly at the beginning of exposure to the up or down conditioning mode (Wolpaw, O'Keefe, Kieffer and Sanders, 1985). Under the control mode, when reflex amplitude does not affect reward probability, descending influence over reflex amplitude is free to vary over its available range. This spontaneous variation presumably contributes to the observed day-to-day variation. As soon as the reward contingency is imposed, the animal is rewarded for maintaining descending influence at the task-appropriate end of its available range (e.g., the end that makes the reflex larger if reward is being given for larger reflexes). This operantly conditioned change in descending influence could account for both the substantial phase I change in reflex amplitude and also for the drop in day-to-day variation. The subsequent gradual phase II change in reflex amplitude suggested that the change in descending influence, as it continued over days and weeks, gradually changed the spinal cord. Subsequent studies sought to detect and define this hypothesized spinal cord plasticity.

5.5 H-REFLEX CONDITIONING CHANGES THE SPINAL CORD

The data illustrated in Figure 5.2 show that operant conditioning of the H-reflex gradually changes spinal cord function, but it does not reveal whether conditioning produces persistent change in the spinal cord, i.e., change that is not continuously dependent on maintenance of appropriate descending influence. To address this question, nerve volley responses were recorded under anesthesia from monkeys after conditioning had increased (HR↑ animals) or decreased (HR↓ animals) the triceps surae H-reflex of one leg (Wolpaw and Lee, 1989). The reflex asymmetry produced by conditioning persisted for at least 2-3 days after spinal cord transection had abolished all descending influence (Figure 5.4A). This finding showed that conditioning had produced activity-driven plasticity in the spinal cord itself. The plasticity, induced by altered descending influence, appeared to be a physiological analog of the plasticity produced by cerebellar damage in the DiGiorgio protocol ((DiGiorgio, 1929; Gerard, 1961; Kandel and Spencer, 1968) and this volume). The difference was that the spinal cord plasticity in the DiGiorgio protocol was induced by a pathological alteration in descending influence (i.e., its origin was a supraspinal lesion) while the spinal cord plasticity in the H-reflex conditioning protocol was induced by an operantly conditioned alteration in descending influence, that is, in the course of learning a motor task.

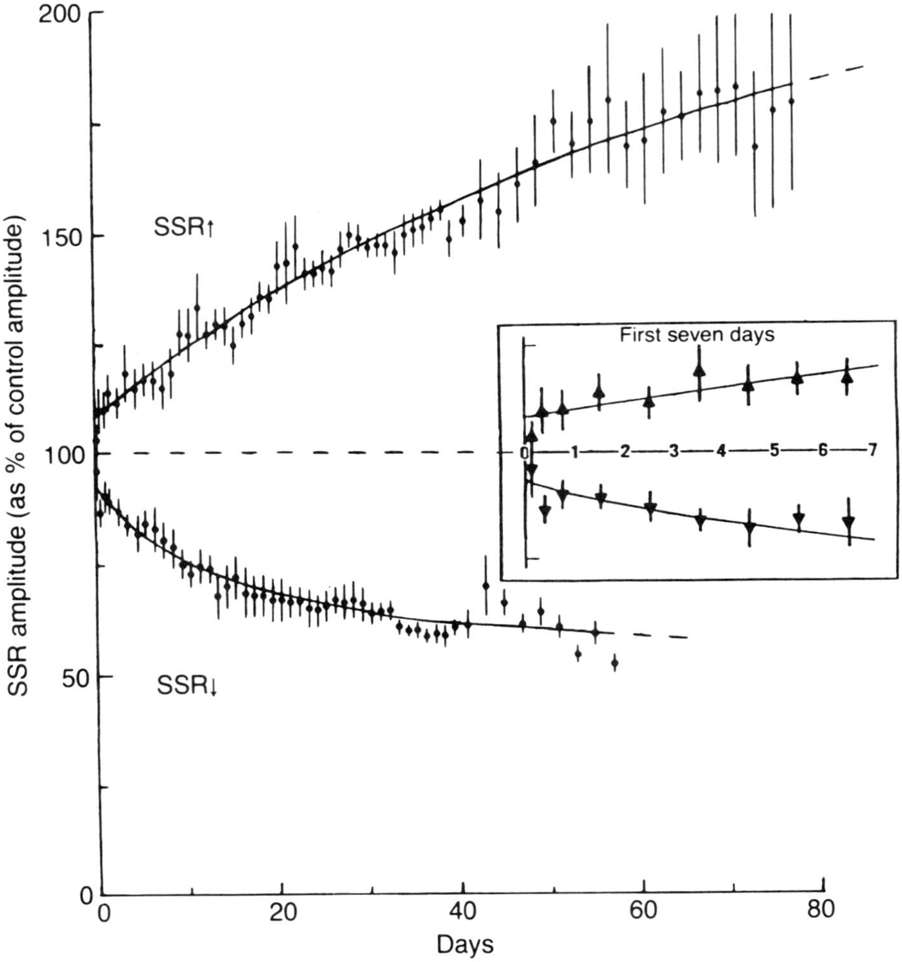

Figure 5.3. Two-phase course of conditioning of the monkey biceps SSR. Average SSR amplitude under the SSR↑ or SSR↓ conditioning mode for 18 mode exposures (11 up and 7 down) in 9 animals. Each value is SSR amplitude (±SEM) as percent of initial amplitude. Values are displayed for hrs 0-6, 6-12, and 12-24 of the first day, for each subsequent day up to 40 days, and for every two days from then on. The inset shows a horizontally-expanded view of the first week. The superimposed lines are plots of the equation "$y = (a/(b+(1/x))) + c$." (Wolpaw and O'Keefe (1984) for details of calculations.) Both the SSR↑ and SSR↓ data plots show a nearly immediate (within 6 hr) mode-appropriate offset (i.e., phase I change), followed by gradual change (i.e., phase II change) that continues for 40 days or more. (Modified from Wolpaw and O'Keefe (1984).)

The nerve volley study indicated that H-reflex conditioning had another effect, in addition to the persistent reflex asymmetry. This second effect was wholly unexpected. Responses on the contralateral (i.e., unconditioned) side were much larger in HR↓ animals than in HR↑ animals (e.g., Figure 5.4A). This difference was seen only in the acute preparations; it had not been evident in the contralateral H-reflexes of the awake behaving animals (Wolpaw, Lee and Calaitges, 1989; Wolpaw, Herchenroder and Carp, 1993). Thus, anesthesia and cord transection revealed a hidden effect of conditioning: conditioning modified the other side of the spinal cord. The possible etiology of this contralateral plasticity is discussed below.

5.6 THE PLASTICITY UNDERLYING H-REFLEX CONDITIONING

Physiological and anatomical studies have begun to define the spinal cord plasticity that accompanies H-reflex conditioning. The initial goal was simply to find the plasticity responsible for the behavioral change, i.e., a larger or smaller H-reflex. That the problem is more complicated was suggested by the several effects revealed by the nerve volley study and was confirmed by subsequent studies.

Intracellular recordings from conditioned animals assessed motoneuron intrinsic properties and the EPSPs produced in motoneurons by homonymous and synergist group I monosynaptic input (Carp and Wolpaw, 1994; Carp and Wolpaw, 1995, Chen, Chen, Sheikh and Wolpaw, 2001). Motoneuron axonal conduction velocity, the property that might be thought least likely to change with conditioning, was in fact lower in motoneurons from the conditioned side of HR↓ animals. In addition, motoneuron firing threshold was shifted positively. As Figure 5.4B shows, the threshold change, combined with the moderate decrease noted in Ia EPSP amplitude, could explain the H-reflex decrease. The positive shift in firing threshold, if it occurs in the axon as well as in the cell body, could also explain the lower motoneuron conduction velocity. A modeling study supported this hypothesis (Halter, Carp and Wolpaw, 1995). It showed that the changes in threshold and conduction velocity could result from nearly identical positive shifts in the activation voltage of Na^+ channels in the motoneuron membrane. Altered activation of protein kinase C might produce such a change in Na^+ channels. How altered descending activity might affect protein kinase C in the motoneuron is unknown at present.

In contrast to the intracellular data from HR↓ animals, the data from HR↑ animals provided no explanation for conditioned H-reflex increase (Carp and Wolpaw, 1995). In fact, the data suggested that H-reflexes on the conditioned side of HR↑ animals should have been slightly below normal. These results suggest that HR↓ and HR↑ conditioning are not mirror images of each other, but instead have different mechanisms. This is consistent with the finding that H-reflex change under the HR↑ or HR↓ mode is not affected by earlier exposure to the opposite mode (Wolpaw, 1983; Chen and Wolpaw, 1996). Thus, for example, the effect of HR↓ mode exposure on an animal's control, or naive, H-reflex is similar in rate and magnitude to its effect on an H-reflex that has been increased by prior exposure to the HR↑mode. The intracellular results from HR↑animals also suggest that the mechanism of HR↑ conditioning might reside outside the two-neuron monosynaptic pathway. Group Ia or Ib excitation elicited by the nerve stimulation could reach the motoneuron via disynaptic pathways in time to affect the H-reflex (Baldissera, Hultborn and Illert, 1981; Burke, Gandevia and McKeon, 1984; Fournier, Meunier, Pierrot-Deseilligny and Shindo, 1986; Jankowska, 1992). Oligosynaptic excitation or inhibition of this kind could add to or subtract from, respectively, the monosynaptic Ia EPSP and thus modify H-reflex amplitude.

In theory, plasticity at the Ia synapse might be considered the most plausible mechanism of H-reflex change (Wolpaw and Carp, 1990). Alterations in presynaptic inhibition at this synapse certainly play a role in short-term task-dependent changes in the functioning of the reflex arc (Baldissera, Hultborn and Illert, 1981; Capaday and Stein, 1987; Stein, 1995; Faist, Dietz and Pierrot-Deseilligny, 1996). Thus, for example, soleus H-reflex amplitude in humans is smaller during walking than during standing and still smaller during running. Because these H-reflex differences are seen even if background EMG activity, which reflects motoneuron tone, is the same, they are attributed to differences in presynaptic inhibition. Furthermore, alteration in presynaptic inhibition would seem to be a sharply focused mechanism for modifying H-reflex amplitude without greatly affecting other aspects of motoneuron function. In view of these considerations, the lack of major changes in the homonymous EPSPs of motoneurons from the conditioned sides of either HR↑ or HR↓ animals was surprising. At the same time, though homonymous EPSPs showed only modest change, heteronymous EPSPs (i.e., those elicited by non-triceps surae Ia afferents) were greatly reduced in both HR↑ and HR↓ animals (Carp and Wolpaw, 1995). Thus, conditioning did seem to affect the Ia synapse, but not as expected.

Electron microscopic examination of motoneuron synaptic coverage defined additional features of the spinal cord plasticity caused by H-reflex conditioning (Feng-Chen and Wolpaw, 1996). F terminals (i.e., terminals that have flattened vesicles, contain GABA and/or glycine, and are thought to be mainly inhibitory (Peters, Palay and Webster, 1990)) on cell bodies and their proximal dendrites were smaller on the conditioned side of HR↑ animals than on the conditioned side of HR↓ animals. F terminals from the unconditioned sides of HR↑ and HR↓ animals and from naive animals were intermediate in size. This difference between HR↑ and HR↓ animals was accompanied on the cell bodies by a difference in the number of active zones per F terminal (Figure 5.4C). This finding implies a comparable difference in synaptic strength. Furthermore, C terminals (i.e., large terminals with subsynaptic cisterns accompanied by rough endoplasmic reticulum (Yamamoto, Hertzberg and Nagy, 1991; Starr and Wolpaw, 1994)) were found to be smaller and more clustered on motoneurons from the conditioned side of HR↑ animals. While the functional effects of the C terminal difference are unknown, smaller F terminals could conceivably be the anatomical basis of decreased disynaptic group I inhibition. Thus, they could be responsible for the H-reflex increase caused by HR↑ conditioning.

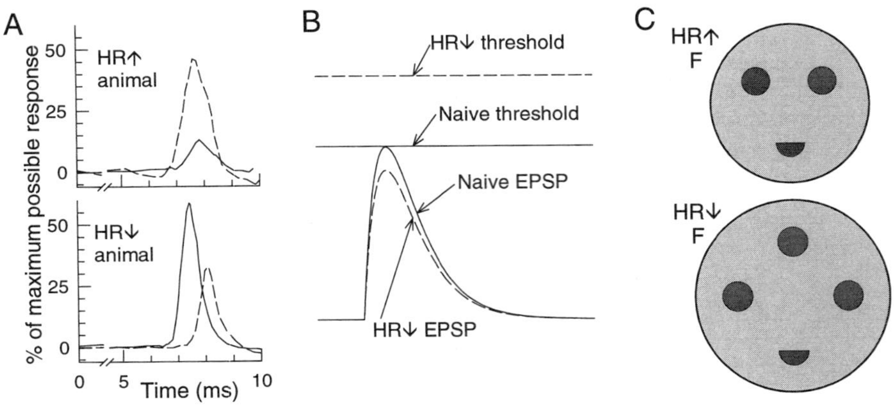

Figure 5.4. Spinal cord plasticity accompanying H-reflex conditioning. (A) Monkeys in which the triceps surae H-reflex in one leg had been increased (HR↑ animal) or decreased (HR↓ animal) by conditioning were anesthetized and surgically prepared, and triceps surae nerve volley responses to stimulation of the proximal ends of the cut dorsal roots were measured under continued anesthesia before and for 2-3 days after thoracic cord transection abolished descending influence. The reflex asymmetry created by conditioning was still evident after transection: nerve volley responses on the conditioned side (dashed lines) were greater than those on the unconditioned side (solid lines) in HR↑ animals and smaller in HR↓ animals. In addition, the responses on the unconditioned side were considerably larger in HR↓ animals than in HR↑ animals. This second finding was not expected because, in the awake behaving animals, H-reflexes on the unconditioned side changed little during conditioning. It showed that conditioning had effects on the contralateral spinal cord that affected behavior only when anesthesia and transection removed descending influence. (B) Triceps surae motoneurons on the conditioned side of HR↓ monkeys were found to have more positive firing thresholds and slightly smaller Ia EPSPs. Together, these two findings could explain why HR↓ motoneurons were less likely to fire in response to nerve stimulation. (C) The contacts of idealized average F terminals and their active zones on the cell bodies of triceps surae motoneurons on the conditioned sides of HR↑ and HR↓ monkeys. HR↑ F terminals are smaller and have fewer active zones. Active zone size is not different. The diameter of the HR↓ F terminal is 2.2 μm. A from Wolpaw and Lee (1989), B from Carp and Wolpaw (1994), C from Feng-Chen and Wolpaw (1996).

5.7 DESCENDING PATHWAYS AND H-REFLEX CONDITIONING

The impetus for H-reflex conditioning is reward delivery, and this event is detected supraspinally. Thus, H-reflex conditioning and the spinal cord plasticity that accompanies it must result from appropriate change in descending influence from supraspinal structures. In both humans and rats, conditioning can still occur after partial spinal cord injury (Segal and Wolf, 1994; Chen, Wolpaw, Jakeman and Stokes, 1996 and 1999). Nevertheless, such injury often impairs conditioning and the degree of impairment correlates with the severity of the injury as assessed by tissue loss. Furthermore, as illustrated in Figure 5.5, recent rat studies have shown that transection of the dorsal column, which in rats includes the main corticospinal tract, prevents HR↓ conditioning (Chen and Wolpaw, 1997); and preliminary data indicate that the corticospinal tract, rather than the dorsal column ascending tract, is the crucial pathway (Chen, Chen and Wolpaw, 2000). In contrast, transection of the entire ipsilateral lateral column, which includes rubrospinal, vestibulospinal, and reticulospinal tracts, does not prevent it (Chen and Wolpaw, 1997). These results are consistent with human studies indicating that strokes damaging motor cortex and associated subcortical structures usually prevent down conditioning of the biceps SSR (Segal, 1997). In combination with the observation that SSR and H-reflex conditioning is specific to the muscle being conditioned (i.e., reflex change is focused in the muscle that determines reward even if reflexes are elicited simultaneously from other muscles (Wolpaw, Seegal and O'Keefe, 1983; Wolpaw, Lee and Calaitges, 1989; Wolpaw, Herchenroder and Carp, 1993; Wolf and Segal, 1996)), these results strongly suggest that the main corticospinal tract is essential for HR↓ conditioning. Its importance for HR↑ conditioning is currently under study.

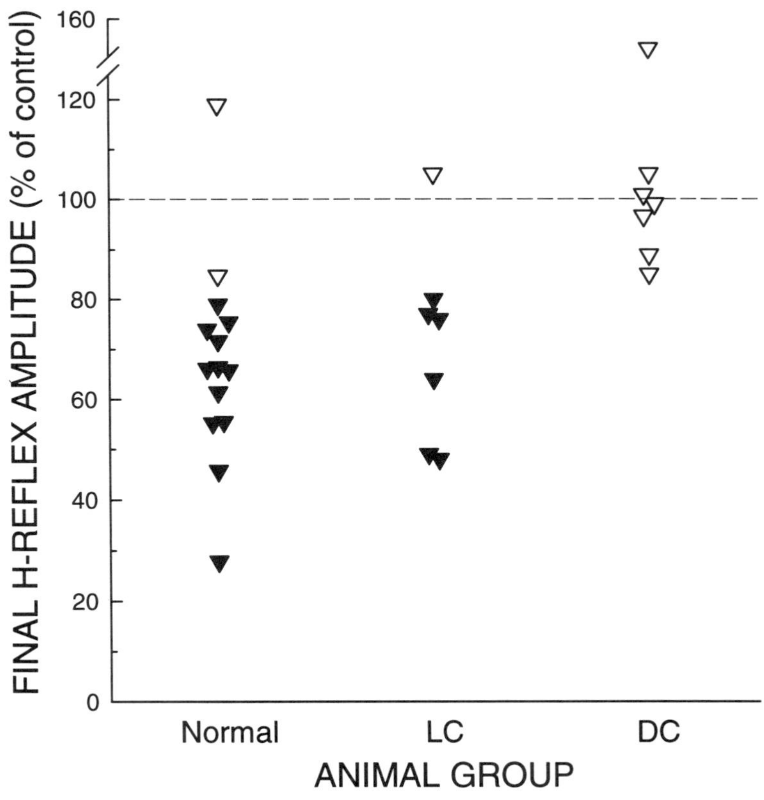

Figure 5.5. Effects of spinal cord lesions on H-reflex conditioning. Final H-reflex values (in percent of initial values) for normal, right lateral column-transected (LC), and dorsal column-transected (DC) rats that were exposed to HRdown conditioning of the right soleus muscle. Successful conditioning (i.e., decrease to ≤80%), shown by filled symbols, occurred in 86% of normal or LC rats, but not in any DC rats. From Chen and Wolpaw (1997) plus additional data.

5.8 PRESENT KNOWLEDGE OF THE NATURE AND ORIGIN OF THE SPINAL CORD PLASTICITY ASSOCIATED WITH H-REFLEX CONDITIONING

Figure 5.6 summarizes current knowledge of the ipsilateral plasticity accompanying H-reflex conditioning. It does not incorporate the currently undefined contralateral plasticity (Wolpaw and Lee, 1989). The figure is based on data from monkeys, rats and humans. It should be recognized that the plasticity could differ considerably among species.

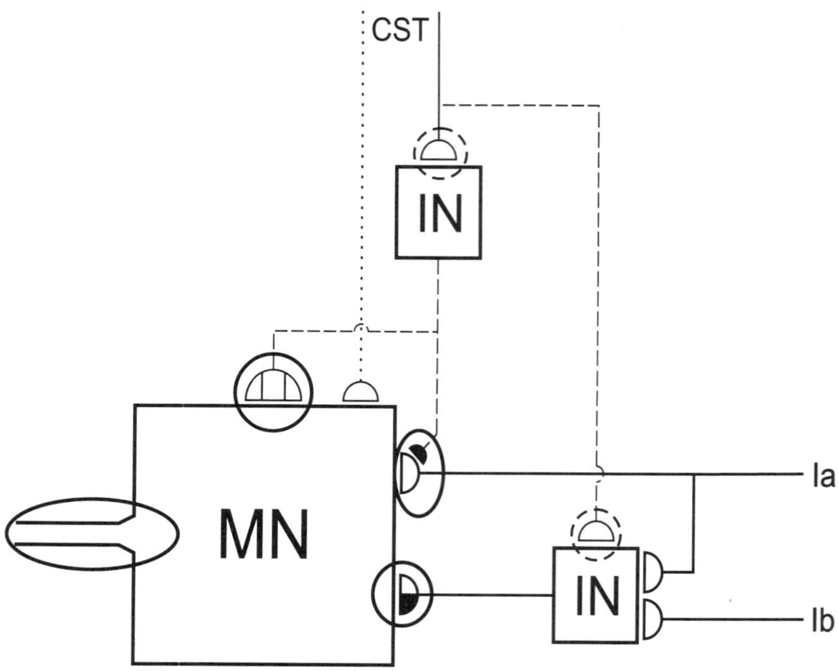

Figure 5.6. Likely locations of spinal cord plasticity and altered descending influence. "MN" indicates the triceps surae motoneuron population, and each "IN" indicates one or more spinal interneuron populations. Open synaptic terminals are excitatory, solid terminals are inhibitory, and half-open terminals could be either. The subdivided terminal is a cluster of C terminals. Dashed pathways imply the possibility of intervening spinal interneurons, and the dotted pathway is uncertain. The monosynaptic and possibly disynaptic H-reflex pathway, originating principally with group Ia and Ib afferents, is shown, and the hypothesized sites of operantly conditioned plasticity in it are circled with solid lines. Beginning at the left and going clockwise around the motoneuron, they are: the motoneuron membrane (i.e., firing threshold of the initial segment and the rest of the axon), C terminals contacting the motoneuron, the Ia afferent synaptic connection, and terminals bringing disynaptic group I inhibition or excitation to the motoneuron. The corticospinal tract (CST) is shown, and the hypothesized sites of the descending influence responsible for the plasticity in the H-reflex pathway are circled with dashed lines. They are: connections on interneurons conveying presynaptic inhibition to the Ia synapse and/or on interneurons supplying C terminals to the motoneuron, and connections on interneurons that carry disynaptic group I inhibition or excitation to the motoneuron.

The available data indicate plasticity at four locations in or near the motoneuron. A positive shift in the firing threshold of the initial segment and the rest of the axon (and presumably the rest of the motoneuron) has been revealed by the intracellular data, is in theory attributable to a positive shift in Na^+ channel activation voltage, and could conceivably account for H-reflex decrease caused by the HR↓ mode (i.e., Figure 4B) (Carp and Wolpaw, 1994; Halter, Carp and Wolpaw, 1995). A modification in disynaptic group I input has been suggested by the anatomical data (i.e., Figure 5.4C) (Feng-Chen and Wolpaw, 1996) (as well as by the negative intracellular physiological data from HR↑ animals (Carp and Wolpaw, 1995)), and could conceivably account for H-reflex increase caused by the HR↑ mode. Anatomical data also suggest a change in the size and grouping of C terminals, which are thought to come from interneurons (Yamamoto, Hertzberg and Nagy, 1991; Feng-Chen and Wolpaw, 1996). Alteration in the Ia afferent synapse is suggested by the modest decrease in homonymous EPSPs on the conditioned side of HR↓ animals and the larger decreases in heteronymous EPSPs in both HR↑ and HR↓ animals (Carp and Wolpaw, 1994; Carp and Wolpaw, 1995), and also by the likelihood that presynaptic inhibition at this connection contributes to short-term task-dependent functional changes (Capaday and Stein, 1987; Stein, 1995; Faist, Dietz and Pierrot-Deseilligny, 1996).

Figure 5.6 also includes the highly probable role of the corticospinal tract in conditioning (Chen and Wolpaw, 1997). The most likely sites of its effect are spinal interneurons that are part of the reflex pathway or affect it directly or through other interneurons. While direct corticospinal tract connections to motoneurons could contribute, these connections are believed to be mainly excitatory (Babalian, Liang and Rouiller, 1993; Giuffrida and Rustioni, 1989; Valtschanoff, Weinberg and Rustioni, 1993) and available anatomical data (Feng-Chen and Wolpaw, 1996) give no evidence for modification in such terminals (i.e., F terminals are inhibitory and C terminals are probably of interneuronal origin). Furthermore, it is not clear whether direct corticomotoneuronal connections to the lumbar cord are present in the rat.

5.9 THE REASONS FOR MULTI-SITE SPINAL CORD PLASTICITY

As Figures 5.4 and 5.6 indicate, we have at present only limited understanding of why the H-reflex is decreased after exposure to the HR↓ mode and increased after exposure to the HR↑ mode. Nevertheless, we know that H-reflex conditioning is accompanied by plasticity at multiple sites in the spinal cord, including sites (e.g., on the contralateral side) that probably do not contribute to the larger or smaller ipsilateral H-reflex (Wolpaw and Lee, 1989; Carp and Wolpaw, 1994; Carp and Wolpaw, 1995; Feng-Chen and Wolpaw, 1996). This complex plasticity is similar to that currently being described in other simple learning models. Plasticity is found at more than one site with conditioning of siphon withdrawal in Aplysia, eyelid closure in rabbits, and vestibuloocular reflex gain in primates (Frost, Clark and Kandel, 1988; Broussard, Bronte-Stewart and Lisberger, 1992; Perrett, Ruiz and

Mauk, 1993; Frost and Kandel, 1995). It seems clear that behavioral changes, however simple, do not result from single changes at single sites.

Although it is consistent with neural network-based concepts of CNS function, such complex plasticity was generally not anticipated. The assumption was that simple learning would be found to have simple mechanisms, that single changes at single locations would prove to explain changes in simple behaviors. In reality, complex plasticity seems to be both necessary and inevitable, for even the simplest learning phenomena.

The necessity for complex plasticity

Any learning, even that as simple as a larger or smaller H-reflex, presents the nervous system with an intricate problem: to change a specific behavior while preserving other behaviors. The CNS handles many tasks, and each of its components participates in many behaviors. This is especially true for motoneurons and other neurons of the spinal cord, which is the final assembly point for behavior. Thus, the plasticity underlying a rewarded behavior, which might be called "primary plasticity," is likely to disturb other behaviors. For H-reflex conditioning, the shift in motoneuron firing threshold seen in HR↓ animals could be primary plasticity: it might help decrease the H-reflex and thereby increase reward probability. At the same time, the threshold shift probably has additional effects, such as reducing motoneuron excitation by excitatory inputs during walking or climbing. Because it disturbs other important behaviors, the threshold change probably induces further learning that preserves or restores them. The additional plasticity responsible for this further learning could be called "compensatory." Compensatory plasticity, like primary plasticity, is recognized through its effects on behavior.

The contralateral spinal cord plasticity revealed by the nerve volley data (Wolpaw and Lee, 1989) (e.g., Figure 5.4A) might be compensatory. It is apparent only when anesthesia and transection have isolated the spinal cord. In the awake behaving animal, supraspinal influence and spinal cord function combine so that behavior, i.e., the H-reflex on the unconditioned side, changes little in the awake behaving animal during conditioning. Nevertheless, as the nerve volley results show, the CNS substrates responsible for this unchanged behavior are not the same after conditioning. The behavior is the same, but its neuronal substrates are different.

The inevitability of complex plasticity

Because the capacity for activity-driven plasticity is nearly ubiquitous in the CNS, plasticity at one location, by changing the activity reaching other locations, is likely to cause further plasticity, which could be labeled "reactive." Like compensatory plasticity, reactive plasticity might have no clear relationship to the newly acquired behavior. Indeed, when analyzed only in the context of that behavior, it might even appear to be maladaptive. Furthermore, the likelihood that some plasticity that in itself improves behavior also leads to reactive plasticity that degrades behavior is likely to help determine the locations and nature of primary and compensatory plasticity.

One recently described example of reactive plasticity may be relevant to H-reflex conditioning. Long-term suppression of activity in Ia afferents (caused by tetrodotoxin-filled nerve cuffs) increases the size of Ia EPSPs in the motoneuron: with long-term reduction in activity, the Ia afferent-motoneuron synapse becomes stronger (Webb and Cope, 1992). This finding suggests that, while increased (HR↓ mode) or decreased (HR↑ mode) presynaptic inhibition at the Ia terminal could produce short-term decrease or increase in the H-reflex, it probably does not account for the gradual decrease (HR↓ mode) or increase (HR↑ mode) that takes place over the weeks of conditioning. If the Ia synapse reacts to suppression by presynaptic inhibition as it does to suppression by tetrodotoxin, change in presynaptic inhibition would not be an effective long-term response to the conditioning task. That other mechanisms underlie the gradual H-reflex change is suggested by the fact that change in Ia EPSPs alone can explain only a small part of the nerve volley asymmetry seen in anesthetized transected animals (Figures 5.4A and B) (Wolpaw and Lee, 1989; Carp and Wolpaw, 1994).

5.10 DEFINING THE TRANSLATION OF PLASTICITY INTO BEHAVIOR

The behavioral effects of a few of the spinal cord changes that accompany H-reflex conditioning (e.g., the threshold change in HR↓ motoneurons) can be predicted with some assurance. However, the effects of other changes, and the causal connections between changes, remain mysterious. For example, the modified F terminals and/or C terminals might convey to the motoneuron the descending influence that generates the threshold shift. Or the altered F terminals might change disynaptic group I inhibition that affects the H-reflex (Carp and Wolpaw, 1995). On the other hand, these synaptic alterations might represent compensatory or reactive plasticity.

Answering these questions, and explaining why the H-reflex gets larger or smaller, depends on fuller description of the complex pattern of spinal cord plasticity that accompanies conditioning, including the contralateral plasticity. It also depends on defining the interaction during performance between descending influence and spinal cord plasticity. This descending influence probably changes over the course of conditioning. Otherwise, the contralateral plasticity found in the nerve volley study (Wolpaw and Lee, 1989) (e.g., Figure 5.4A) would substantially modify the contralateral H-reflex (which does not happen), and the shift in motoneuron firing threshold (Figure 5.4B) would be likely to lead to a fall in background EMG (which also does not happen). For instance, a rise in tonic descending excitation might counterbalance the effect of the threshold shift on tonic motoneuron activity (without necessarily canceling the effect of the shift on the H-reflex), and thereby preserve the level of background EMG.

At least some aspects of the spinal cord plasticity under study directly underlie the H-reflex change. At the same time, because H-reflex change is motivated by an alimentary reward and because it results from long-term change in descending influence, it is likely to involve supraspinal plasticity also. The changes in somatosensory evoked potentials that accompany H-reflex conditioning suggest that plasticity occurs in sensorimotor cortex (Wolpaw and Dowman, 1988), and this possibility is supported by the considerable evidence of such plasticity in numerous other situations (e.g., ablations, specific stimulation paradigms) (Merzenich and Jenkins, 1993; O'Leary, Ruff and Dyck, 1994). Cortical changes can occur quickly, and might be responsible for the change in descending influence that gradually modifies the spinal cord. Cerebellar plasticity seems probable from results in other conditioning models (Raymond, Lisberger and Mauk, 1996) and from the effects of cerebellar stimulation (Ebner, Bloedel, Vitek and Schwartz, 1982; Upton and Cooper, 1976). Further delineation of the complex spinal cord plasticity and the descending influence that creates it should help guide study of the accompanying supraspinal plasticity.

The central goal is to delineate all the major elements of the plasticity caused by the H-reflex conditioning protocol and to understand their impacts on behavior. Because phenomena similar to H-reflex conditioning play a role in normal development and learning (see below), realization of this objective should help in understanding the acquisition and maintenance of more complex motor skills.

5.11 CONDITIONING OF THE SSR PATHWAY AS A PART OF SKILL ACQUISITION

The experimental protocol illustrated in Figure 5.2 uses the H-reflex, or the SSR itself, as a simple behavior, and H-reflex conditioning as a simple motor skill. Outside the laboratory, the SSR is normally a part of much more complicated behaviors. Monosynaptic Ia afferent excitation of motoneurons contributes to behaviors from posture and locomotion to the most difficult athletic and technical performances. Evidence of several kinds indicates that SSR plasticity similar to that produced in the laboratory contributes to early motor development and to the acquisition of motor skills throughout life, and that the plasticity is induced and guided by activity in descending spinal cord pathways.

In newborn infants, muscle stretch elicits SSRs from both agonist and antagonist muscles (Myklebust, Gottlieb and Agarwal, 1986; O'Sullivan, Eyre and Miller, 1991). The antagonist SSRs disappear in childhood, except when perinatal supraspinal damage (e.g., cerebral palsy) disrupts descending influence (Myklebust, Gottlieb, Penn and Agarwal, 1982). When normal descending activity is absent, antagonist SSRs can last into adulthood, and can contribute to motor disabilities. Figure 5.7A contrasts agonist and antagonist SSRs from a normal infant, a normal adult, and an adult with cerebral palsy.

In normal adults, SSRs and H-reflexes reflect the nature, intensity, and duration of past motor training. When monkeys trained to make smooth forearm movements were exposed to random brief perturbations of those movements, and when this exposure continued over months, the SSRs produced by the perturbations gradually increased, and the increase appeared to account for improvement in performance (Figure 5.7B) (Meyer-Lohmann, Christakos and Wolf, 1986). Longer latency responses decreased as the SSR took over the task of opposing the perturbations, and the deflections caused by the perturbations became smaller. In professional ballet dancers, leg H-reflexes and SSRs (and group Ia reciprocal inhibition as well) are considerably smaller than in other highly trained athletes (Figure 5.7C) (Koceja, Burke and Kamen, 1991; Nielsen, Crone and Hultborn, 1993). The decrease in monosynaptic peripheral influence on motoneurons indicated by the smaller reflexes might conceivably serve to increase cortical control and thereby allow more precisely controlled movements. If this is true, then the gradual decrease in these spinal reflexes is part of the process of skill acquisition that occurs over months and years of training, and helps to explain why performance improves with such prolonged practice. These examples suggest that the conditioning of the SSR pathway studied in the laboratory represents a phenomenon that has a significant role in acquisition and maintenance of motor skills throughout life.

5.12 THE CONTRIBUTION OF SPINAL CORD PLASTICITY TO LEARNING

Data like that summarized in the preceding section, together with SSR and H-reflex conditioning and with the chronic effects of supraspinal trauma on spinal cord function, show that the intrinsic properties and connectivities of spinal cord motoneurons and their associated interneurons reflect the long-term influences of descending pathways. Because these spinal cord neurons comprise the final common pathway for all motor behavior, long-term descending influences must reflect the requirements of numerous behaviors, and, as a result, the properties and connectivities of spinal cord neurons at any point in time are a compromise that serves many different masters (Wolpaw and Tennissen, 2001). For example, if the animal that supplied the data shown in Figure 5.7B had been trained on alternate days on two different tasks, one that rewarded greater opposition to the perturbation and another that rewarded less opposition, the SSR would probably not have increased greatly over training. but would rather have approached some intermediate amplitude that represented a compromise between the conflicting demands of the two tasks.

The slow rate of spinal cord plasticity, illustrated in Figure 5.2, can be accounted for by its dependence on continuously shifting combinations of ongoing descending influences. Even if one of these influences changes abruptly, compensatory or reactive changes in other influences are likely to ensure that any changes in spinal cord function occur gradually. On the other hand, the mechanisms underlying spinal cord plasticity may or may not be themselves slow. For example, long-term potentiation occurs in the spinal cord (Randic, Jiang and Cerne, 1993; Pockett, 1995; Lozier and Kendig, 1995; Liu and Sandkuhler, 1997), and might contribute to the plasticity underlying behavioral changes such as H-reflex conditioning. The continuing interactions between multiple rapid changes could produce the gradual changes seen at the behavioral level.

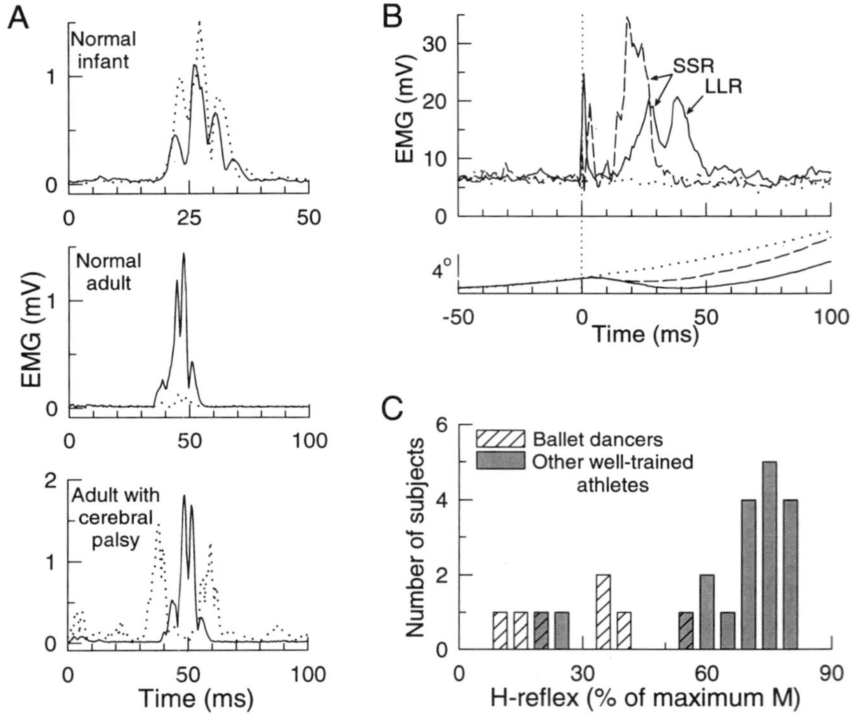

Figure 5.7. SSR pathway conditioning during development and during acquisition of motor skills. (A) EMG responses of human soleus (solid) and tibialis anterior (dotted) muscles to sudden foot dorsiflexion, which stretches the soleus and shortens its antagonist, the tibialis anterior. In a normal infant, this stretch elicits SSRs from both muscles. In a normal adult, an SSR is elicited only in the stretched muscle, i.e., the soleus. Little or no response is seen in the tibialis anterior. However, in an adult with cerebral palsy, in whom perinatal supraspinal injury has impaired the long-term descending activity responsible for development of normal adult SSRs, the infantile pattern remains: soleus stretch elicits SSRs in both soleus and tibialis anterior. (The short latency and unusual form of the tibialis anterior response also suggest central and/or peripheral abnormalities.) (B) Working for reward, monkeys performed an elbow flexion-extension task. Brief perturbations (i.e., 10-ms torque pulses) were randomly superimposed. Displayed are biceps EMG and elbow angle (flexion is upward) for an unperturbed trial (dotted), a perturbed trial early in training (solid), and a perturbed trial later in training (dashed). Early in training, perturbation produces both an SSR and a long-latency polysynaptic response (i.e., LLR). After intermittent training over several years, the SSR has greatly increased and the LLR has disappeared. The SSR has gradually taken over the task of counteracting the perturbation. This is associated with improved performance: the deflection imposed by the torque pulse on the otherwise smooth course of elbow flexion is smaller and briefer. (C) Soleus H reflexes are far smaller in professional ballet dancers than in other well-trained athletes (e.g., runners, swimmers, cyclists). H-reflexes of sedentary subjects are in between. A from B. Myklebust (unpublished data similar to those in Myklebust, Gottlieb, and Agarwal (1986) and Myklebust, Gottlieb, Penn, and Agarwal (1982)), B from Meyer-Lohmann, Christakos, and Wolf (1986), C from Nielsen, Crone, and Hultborn (1993).

The spinal cord neuronal properties and connectivities defined over time by multiple descending influences provide an apparatus for the execution of all of an individual's roster of motor performances. Thus, the wide variations in spinal reflexes commonly found among individuals (e.g., Figure 5.7C) are presumably due in considerable part to differences in past training, that is, in the requirements placed on the motor system, and to the fact that different nervous systems confronting similar tasks may arrive at different solutions.

5.13 SPINAL CORD PLASTICITY AND PLASTICITY ELSEWHERE IN THE CNS

Spinal cord plasticity, which is induced, maintained, and adjusted by descending influence and provides an accommodating substrate for the execution of motor performances, appears to serve a purpose similar to the plasticity elsewhere in the CNS that supports language and other high-level skills such as mathematics. These skills also depend on prolonged learning processes that gradually affect neurons and their connectivities. For example, language acquisition entails and depends on the development of appropriate sensitivities to incoming sounds and the development of appropriate ranges of sound production. Like the changes in spinal reflexes illustrated in Figure 5.7A, this plasticity is particularly prominent early in life (e.g., Kuhl, 1998). While languages can be learned, and spinal reflexes can be modified, later in life, this learning is clearly constrained by the patterns established in the first few years. A language acquired later on is usually spoken with an accent derived from the individual's original language, and reflex conditioning later in life probably cannot reestablish the antagonist excitation lost early in a normal childhood.

Gradually acquired plasticity, in the spinal cord and elsewhere, is an important element in the proper execution of many, probably most, behaviors. As a result, understanding the acquisition and maintenance of motor skills and other complex behaviors necessarily requires investigation of this plasticity. The spinal cord is a logical place to begin this study, first because of its central role in all behaviors, both simple and complex, and second because it is relatively simple and accessible. Elucidation of the location and nature of spinal cord plasticity should facilitate and guide study of the supraspinal mechanisms that create it and the supraspinal plasticity, in cortex and elsewhere, that combines with it to produce skilled performances.

ACKNOWLEDGMENTS

Drs. Jonathan S. Carp, Xiang Yang Chen, and Dennis J. McFarland provided valuable comments on the manuscript. Ms. Dayna Maniccia Stark provided technical assistance. Work in the author's laboratory has been supported by the National Institutes of Health, the United Cerebral Palsy Research and Educational

Foundation, the Paralyzed Veterans of America Spinal Cord Research Foundation, the American Paralysis Association, and the International Spinal Research Trust.

5.14 REFERENCES

Babalian, A., Liang, F. and Rouiller, E.M. Cortical influences on cervical motoneurons in the rat: recordings of synaptic responses from motoneurons and compound action potential from corticospinal axons. *Neuroscience Research* 1993; 16:301-10

Baldissera, F., Hultborn, H. and Illert, M. "Integration in spinal neuronal systems." In *Handbook of Physiology. Section I: The Nervous System. Volume II: Motor Control, Part I.* V B. Brooks, ed. Baltimore: Williams and Wilkins Company, 1981.

Broussard, D.M., Bronte-Stewart, H.M. and Lisberger, S.G. Expression of motor learning in the response of the primate vestibuloocular reflex pathway to electrical stimulation. *Journal of Neurophysiology* 1992; 67:1493-508

Brown, W.F. *The Physiological and Technical Basis of Electromyography.* Boston: Butterworths, 1984

Burke, D, Gandevia, S.C., and McKeon, B. Monosynaptic and oligosynaptic contributions to human ankle jerk and H-reflex. *Journal of Neurophysiology* 1984; 52:435-48

Calabresi, P., Pisani, A., Mercuri, N.B., and Bernardi, G. The corticostriatal projection - from synaptic plasticity to dysfunctions of the basal ganglia. *Trends in Neuroscience,* 1996; 19:19-24

Capaday, C. and Stein, R.B. Difference in the amplitude of the human soleus H reflex during walking and running. *Journal of Physiology* (Lond), 1987; 392:513-22

Carp, J.S., Chen, X.Y., Sheikh, H. and Wolpaw, J.R. Operant conditioning of rat H-reflex affects motoneuron axonal conduction velocity. *Experimental Brain Research,* 2001; 136:269-273.

Carp, J.S. and Wolpaw, J.R. Motoneuron plasticity underlying operantly conditioned decrease in primate H-reflex. *Journal of Neurophysiology,* 1994; 72:431-42

Carp, J.S. and Wolpaw, J.R. Motoneuron properties after operantly conditioned increase in primate H-reflex. *Journal of Neurophysiology,* 1995; 73:1365-73

Chen, X.Y., Chen, L. and Wolpaw, J.R. The corticospinal tract in development and maintenance of H-reflex operant conditioning in rats. *Society for Neuroscience Abstracts,* 2000; 26: 2206.

Chen, X.Y., Chen, L. and Wolpaw, J.R. Time course of H-reflex conditioning in the rat. *Neuroscience Letters,* 2001: in press.

Chen, X.Y. and Wolpaw, J.R. Operant conditioning of H-reflex in freely moving rats. *Journal of Neurophysiology*, 1995; 73:411-5

Chen, X.Y. and Wolpaw, J.R. Reversal of H-reflex operant conditioning in the rat. *Experimental Brain Research,* 1996; 112:58-62

Chen, X.Y. and Wolpaw, J.R. Dorsal column but not lateral column transection prevents down conditioning of H-reflex in rats. *Journal of Neurophysiology,* 1997; 78:1730-4

Chen, X.Y., Wolpaw, J.R., Jakeman, L.B. and Stokes, B.T. Operant conditioning of H-reflex in spinal cord-injured rats. *Journal of Neurotrauma,* 1996; 13:755-66

Chen, X.Y., Wolpaw, J.R., Jakeman, L.B. and Stokes, B.T. Operant conditioning of H-reflex increase in spinal cord-injured rats. *Journal of Neurotrauma,* 1999; 16:175-186

DiGiorgio, A.M. Persistenza nell'animale spinale, di asymmetrie posturali e motorie di origine cerebellare: I, II, III. *Archivio di Fisiologia,* 1929; 27:518-80

Ebner TJ, Bloedel JR, Vitek JL, and Schwartz AB. The effects of cerebellar stimulation on the stretch reflex in the spastic monkey. *Brain,* 1982; 105:425-42

Evatt, M.L., Wolf, S.L. and Segal, R.L. Modification of human spinal stretch reflexes: preliminary studies. *Neuroscience Letters,* 1989; 105:350-5

Faist, M., Dietz, V. and Pierrot-Deseilligny, E. Modulation, probably presynaptic in origin, of monosynaptic Ia excitation during human gait. *Experimental Brain Research,* 1996; 109:441-9

Feng-Chen, K.C, Chen, X.Y. and Wolpaw, J.R. Presynaptic (p) terminals on rat motoneuron primary afferent terminals. *Society for Neuroscience Abstracts,* 1998; 24:1153

Feng-Chen, K.C. and Wolpaw, J.R. Operant conditioning of H-reflex changes synaptic terminals on primate motoneurons. *Proceedings of the National Academy of Sciences USA,* 1996; 93:9206-11

Fisher, S.A, Fischer, T.M. and Carew, T.J. Multiple overlapping processes underlying short-term synaptic enhancement. *Trends in Neuroscience,* 1997; 20:170-7

Fournier, E., Meunier, S., Pierrot-Deseilligny, E. and Shindo, M. Evidence for interneuronally mediated Ia excitatory effects to human quadriceps motoneurons. *Journal of Physiology* (Lond), 1986; 377:143-69

Frost, W.N, Clark, G.A. and Kandel, E.R. Parallel processing of short-term memory for sensitization in Aplysia. *Journal of Neurobiology,* 1988; 19:297-334

Frost, W.N. and Kandel, E.R. Structure of the network mediating siphon-elicited siphon withdrawal in Aplysia. *Journal of Neurophysiology,* 1995; 73:2413-27

Gerard, R.W. "The fixation of experience." In *Brain Mechanisms and Learning,* R W. Gerard, J Konorski, (Eds). Oxford: Blackwell, 1961

Giuffrida, R. and Rustioni, A. Glutamate and aspartate immunoreactivity in corticospinal neurons of rats. *Journal of Comparative Neurology,* 1989; 288:154-64

Halter, J.A., Carp, J.S, and Wolpaw, J.R. Operantly conditioned motoneuron plasticity: possible role of sodium channels. *Journal of Neurophysiology,* 1995; 74:867-71

Henneman, E. and Mendell, L.M. "Functional organization of motoneuron pool and inputs." In *Handbook of Physiology. Sect. I. The Nervous System. Vol. II. Motor Control. Part I,* V B. Brooks, (Ed.) Baltimore: Williams and Wilkins, 1981

Jankowska, E. Interneuronal relay in spinal pathways from proprioceptors. *Progress in Neurobiology,* 1992; 38:335-78

Kandel, E.R. and Spencer, W.A. Cellular neurophysiological approaches in the study of learning. *Physiological Review,* 1968; 48:65-134

Koceja, D.M, Burke, J.R. and Kamen, G. Organization of segmental reflexes in trained dancers. *International Journal of Sports Medicine,* 1991; 12:285-9

Kuhl, P.K. "The Development of Speech and Language." In *Mechanistic Relationships Between Development and Learning,* T J. Carew, R Menzel, and CJ Shatz, (Eds). New York: John Wiley & Sons, Inc. 1998

Liu, X. and Sandkuhler, J. Characterization of long-term potentiation of C-fiber-evoked potentials in spinal dorsal horn of adult rat: essential role of NK1 and NK2 receptors. *Journal of Neurophysiology,* 1997; 78:1973-82

Lozier, A.P. and Kendig, J.J. Long-term potentiation in an isolated peripheral nerve-spinal cord preparation. *Journal of Neurophysiology,* 1995; 74:1001-9

Magladery, J.W., Porter, W.E., Park, A.M. and Teasdall, R.D. Electrophysiological studies of nerve and reflex activity in normal man. IV. The two-neuron reflex and identification of certain action potentials from spinal roots and cord. *Bulletin John Hopkins Hospital,* 1951; 88:499-519

Maren, S. and Baudry, M. Properties and mechanisms of long-term synaptic plasticity in the mammalian brain: relationships to learning and memory. *Neurobiology of Learning and Memory,* 1995; 63:1-18

Matthews, P.B.C. *Mammalian Muscle Receptors and Their Central Actions.* Baltimore: Williams & Wilkins, 1972

Mendell, L.M. Modifiability of spinal synapses. *Physiology Review,* 1984; 64:260-324

Merzenich, M.M. and Jenkins, W.M. Reorganization of cortical representations of the hand following alterations of skin inputs induced by nerve injury, skin island transfers, and experience. *Journal of Hand Therapy,* 1993; 6:89-104

Meyer-Lohmann, J., Christakos, C.N. and Wolf, H. Dominance of the short-latency component in perturbation induced electromyographic responses of long-trained monkeys. *Experimental Brain Research,* 1986; 64:393-9

Muir, G.D. and Steeves, J.D. Sensorimotor stimulation to improve locomotor recovery after spinal cord injury. *Trends in Neuroscience,* 1997; 20:72-7

Myklebust, B.M., Gottlieb, G.L. and Agarwal, G.C. Stretch reflexes of the normal human infant. *Developmental Medicine and Child Neurology,* 1986; 28:440-9

Myklebust, B.M., Gottlieb, G.L., Penn, R.L. and Agarwal, G.C. Reciprocal excitation of antagonistic muscles as a differentiating feature in spasticity. *Annals of Neurology,* 1982; 12:367-74

Nielsen, J., Crone, C. and Hultborn, H. H-reflexes are smaller in dancers from the Royal Danish Ballet than in well-trained athletes. *European Journal of Applied Physiology,* 1993; 66:116-21

O'Leary, D.D., Ruff, N.L. and Dyck, R.H. Development, critical period plasticity, and adult reorganizations of mammalian somatosensory systems. *Current Opinions in Neurobiology,* 1994; 4:535-44

O'Sullivan, M.C., Eyre, J.A, and Miller, S. Radiation of phasic stretch reflex in biceps brachii to muscles of the arm in man and its restriction during development. *Journal of Physiology,* 1991; 439:529-43

Oda, Y., Charpier, S., Murayama, Y., Suma, C. and Korn, H. Long-term potentiation of glycinergic inhibitory synaptic transmission. *Journal of Neurophysiology,* 1995; 74:1056-74

Perrett, S.P., Ruiz, B.P. and Mauk, M.D. Cerebellar cortex lesions disrupt learning-dependent timing of conditioned eyelid responses. *Journal of Neuroscience,* 1993; 13:1708-18

Peters, A,, Palay, S.L. and Webster, H.D. *Fine Structure of the Nervous System*. New York: Oxford University Press, 1990

Pockett, S. Long-term potentiation and depression in the intermediate gray matter of rat spinal cord in vitro. *Neuroscience*, 1995; 67:791-8

Randic, M,, Jiang, M.C. and Cerne, R. Long-term potentiation and long-term depression of primary afferent neurotransmission in the rat spinal cord. *Journal of Neuroscience*, 1993; 13:5228-41

Raymond, J.L., Lisberger, S.G. and Mauk, M.D. The cerebellum: a neuronal learning machine? *Science*, 1996; 272:1126-31

Segal, R.L. Plasticity in the central nervous system: Operant conditioning of the spinal stretch reflex. *Topics in Stroke Rehabilitation*, 1997; 3:76-87

Segal, R.L. and Wolf, S.L. Operant conditioning of spinal stretch reflex in patients with spinal cord injuries. *Experimental Neurology*, 1994; 130:202-13

Starr, K.A. and Wolpaw, J.R. Synaptic terminal coverage of primate triceps surae motoneurons. *Journal of Comparative Neurology*, 1994; 345:345-58

Stein, R.B. Presynaptic inhibition in humans. *Progress Neurobiology*, 1995; 47:533-44

Upton, A.R. and Cooper, I.S. Some neurophysiological effects of cerebellar stimulation in man. *Canada Journal of Neurological Science*, 1976; 3:237-54

Valtschanoff, J.G, Weinberg, R,J. and Rustioni, A. Amino acid immunoreactivity in corticospinal terminals. *Experimental Brain Research*, 1993; 931:95-103

Webb, C.B. and Cope, T.C. Modulation of Ia EPSP amplitude: the effects of chronic synaptic inactivity. *Journal of Neuroscience*, 1992; 12:338-44

Wolf, S.L. and Segal, R.L. Conditioning of the spinal stretch reflex: implications for rehabilitation. *Physical Therapy*, 1990; 70:652-6

Wolf, S.L. and Segal, R.L. Reducing human biceps brachii spinal stretch reflex magnitude. *Journal of Neurophysiology*, 1996; 75:1637-46

Wolf SL, Segal RL, Heter ND, and Catlin PA. Contralateral and long latency effects of human biceps brachii stretch reflex conditioning. *Experimental Brain Research*, 1995; 107:96-102

Wolpaw, J.R. Adaptive plasticity in the primate spinal stretch reflex: Reversal and re-development. *Brain Research*, 1983; 278:299-304

Wolpaw, J.R. Operant conditioning of primate spinal reflexes: the H-reflex. *Journal of Neurophysiology*, 1987; 57:443-58

Wolpaw, J.R., Braitman, D.J. and Seegal, R.F. Adaptive plasticity in the primate spinal stretch reflex: initial development. *Journal of Neurophysiology*, 1983;50:1296-311

Wolpaw, J.R. and Carp, J.S. Memory traces in spinal cord. *Trends in Neuroscience* 1990;13:137-42

Wolpaw, J.R. and Dowman, R. Operant conditioning of primate spinal reflexes: Effects on cortical SEPs. Electroencephalogr. *Clinical Neurophysiology* 1988; 69:398-401

Wolpaw, J.R. and Herchenroder, P.A. Operant conditioning of H-reflex in freely moving monkeys. *Journal of Neuroscience Methods,* 1990; 31:145-52

Wolpaw, J.R., Herchenroder, P.A. and Carp, J.S. Operant conditioning of the primate H-reflex: factors affecting the magnitude of change. *Experimental Brain Research,* 1993; 97:31-9

Wolpaw, J.R. and Lee C.L. Memory traces in primate spinal cord produced by operant conditioning of H-reflex. *Journal of Neurophysiology,* 1989; 61:563-72

Wolpaw, J.R., Lee, C.L. and Calaitges, J.G. Operant conditioning of primate triceps surae H-reflex produces reflex asymmetry. *Experimental Brain Research,* 1989; 75:35-9

Wolpaw, J.R., Maniccia, D.M. and Elia, T. Operant conditioning of the primate H-reflex: course of development. *Neuroscience Letters,* 1994; 170:203-7

Wolpaw, J.R. and O'Keefe, J.A. Adaptive plasticity in the primate spinal stretch reflex: evidence for a two-phase process. *Journal of Neuroscience,* 1984; 4:2718-24

Wolpaw, J.R., O'Keefe, J.A. and Dowman R. Adaptive plasticity in the primate spinal stretch reflex: A two-phase process. *Society for Neuroscience Abstracts,* 1984; 10: 129

Wolpaw, J.R., O'Keefe, J.A., Kieffer, V.A. and Sanders, M.G. Reduced day-to-day variation accompanies adaptive plasticity in the primate spinal stretch reflex. *Neuroscience Letters,* 1985; 54: 165-71

Wolpaw, J.R., Seegal, R.F., and O'Keefe, J.A. Adaptive plasticity in the primate spinal stretch reflex: behavior of synergist and antagonist muscles. *Journal of Neurophysiology,* 1983; 50:1312-9

Wolpaw, J.R. and Tennissen, A.M. Activity-dependent spinal cord plasticity in health and disease. *Annual Review of Neursoscience,* 2001; 24: 807-43.

Yamamoto ,T., Hertzberg, E.L., and Nagy, J.I. Subsurface cisterns in alpha-motoneurons of the rat and cat: immunohistochemical detection with antibodies against connexin32. *Synapse,* 1991; 8: 119-36

6 MECHANISMS OF CENTRAL SENSITIZATION OF NOCICEPTIVE DORSAL HORN NEURONS

William D. Willis, Jr.
Department of Anatomy and Neurosciences and
Marine Biomedical Institute
University of Texas Medical Branch
Galveston, Texas 77555-1069

6.1 INTRODUCTION

An important theme in contemporary research on the nervous system is plasticity. The activity of the nervous system is in part dependent on its past activity, which implies that neural responses can be modified by prior events. The mechanisms that are thought to underlie many forms of learning and memory, such as long-term potentiation (LTP) and long-term depression (LTD), typify this type of plasticity (Riedel, et al., 1996; Thompson, et al., 1997; Daniel, et al., 1998; however, cf. Hölscher, 1997). Plasticity occurs not only in the brain, but also in the spinal cord (Windhorst, 1996; Willis, 1997), as is emphasized in this symposium volume. For example, electrophysiological response changes similar to LTP and LTD have been described in the spinal cord, both *in vivo* and *in vitro* (Randic, et al., 1993; Svendsen, et al., 1997; Liu and Sandkuhler, 1997; Sandkuhler and Liu, 1998), and rewiring of the afferent connections to the dorsal horn has been demonstrated following peripheral nerve injury (Woolf, et al., 1992). Many workers in the field of pain research refer to the form of spinal cord plasticity that will be discussed in this chapter as central sensitization and suggest that this process underlies such abnormal pain states as secondary hyperalgesia and allodynia (Woolf, 1992; Willis,

1993; cf., Kenshalo, et al., 1982; Woolf, 1983; Yaksh, 1989; Dubner and Ruda, 1992).

Strong or prolonged noxious stimuli can cause an increase in the responsiveness of nociceptive neurons to innocuous and/or noxious stimuli, and this change is called sensitization. If the sensitized nociceptive neurons are primary afferent nociceptors, the process is called peripheral sensitization. If the sensitized nociceptive neurons are located in the central nervous system, such as in the spinal cord dorsal horn, the process is called central sensitization. Central sensitization has many features that resemble LTP (Willis, 1997).

In this chapter, evidence will be given concerning the mechanisms underlying the central sensitization of primate spinothalamic tract (STT) neurons that results from the injection of capsaicin into the skin in the receptive fields of the neurons being studied (Dougherty and Willis, 1992; Simone, et al., 1991). There are at least three reasons for adopting this particular model for the study of central sensitization.

First, there have been psychophysical studies of the pain states that are produced in human subjects by intradermal injection of capsaicin (Simone, et al., 1989; LaMotte, et al., 1991; 1992; Torebjörk, et al., 1992). Changes in the responses of STT neurons in monkeys to stimulation of the skin after capsaicin injection parallel remarkably closely the psychophysical responses of humans, suggesting a direct relevance of the animal studies to the human pain experience (Simone, et al., 1991).

Secondly, the duration of the central sensitization of STT neurons when tested with innocuous or weakly noxious stimuli is about 1.5 to 2 hours. In a given experiment, 2 injections of capsaicin can be injected intradermally into different parts of the receptive field several hours apart, and these result in comparable responses (see Figure 6.7, left sets of bars, lower row of panels). Therefore, a control response to capsaicin can be compared to the response to capsaicin following an experimental manipulation for a given neuron (Dougherty, et al., 1994; Lin, et al., 1999a). This makes it possible to do a pharmacological analysis of the neurotransmitters and subsequent intracellular events that underlie central sensitization by administering drugs that affect the response to one of two successive capsaicin injections.

Thirdly, the central sensitization that follows an intradermal injection of capsaicin appears to result from a time-limited burst of activity in capsaicin-sensitive primary afferent fibers (Baumann, et al., 1991; LaMotte, et al., 1992), and so it is not maintained by a continuous peripheral input. This feature makes it easy to distinguish between changes in the responses of central nociceptive neurons that are intrinsic to the spinal cord and changes in the responsiveness of central neurons that are due to continuous afferent activity. The latter may account for part of the enhanced excitation seen in the formalin and neuropathic pain models (see below; Puig and Sorkin, 1996; Yoon, et al., 1996).

6.2 PSYCHOPHYSICAL RESPONSES TO CAPSAICIN INJECTION

A number of sensory alterations follow the injection of capsaicin into human skin, including pain, hyperalgesia and allodynia. Hyperalgesia is defined as "an increased response to a stimulus which is normally painful" and allodynia refers to "pain due to a stimulus which does not normally provoke pain (Merskey and Bogduk, 1994). The adjectives "mechanical" or "thermal" are applied to hyperalgesia and allodnia produced by specific types of stimuli, such as noxious compression of the skin, heat or cold for forms of hyperalgesia and weak tactile, cool or warm stimuli for different types of allodynia.

When capsaicin is injected into the skin of a human subject, the initial sensory event is severe burning pain at the injection site. This is maximal immediately, within 5 to 15 seconds of the injection, but soon decreases to a tolerable level and reaches a low intensity within minutes (Simone, et al., 1989; LaMotte, et al., 1991; 1992; Torebjörk, et al., 1992). The magnitude and duration of pain are dependent on the amount of capsaicin that is injected. For example, a dose of 0.1 μg of capsaicin injected intradermally produces more intense pain than does injection of the vehicle, and higher doses evoke progressively more intense pain. Furthermore, the mean duration of the pain increases from 1.2 to 3.3 to 8.1 minutes for injections of 0.1, 1 and 10 μg of capsaicin into the skin. After a high dose of capsaicin (100 μg) is injected, the pain is judged to be near the maximum conceivable pain. This pain is rated as 2.6 times as painful as a heat stimulus of 51° C with a 5 second duration, and the pain lasts on average 16.7 minutes (Simone, et al., 1989; 1991).

In the skin over the bleb that forms at the capsaicin injection site, an anesthetic zone can be recognized (LaMotte, et al., 1991). This can be attributed to desensitization of capsaicin-sensitive afferent terminals in the skin and their later degeneration (Carpenter and Lynn, 1981; Simone, et al., 1998; Tominaga, et al., 1998). Surrounding the anesthetic zone is a zone of primary heat hyperalgesia (Simone, et al., 1987). Primary hyperalgesia is attributed at least in part to peripheral sensitization of primary afferent nociceptors (Meyer and Campbell, 1981; Raja, et al., 1988). Initial efforts failed to demonstrate sensitization of primary afferent nociceptors to mechanical or heat stimulation following intradermal injection of capsaicin (Baumann, et al., 1991; LaMotte, et al., 1992). However, sensitization of nociceptors could be produced by topical application of capsaicin (LaMotte, et al., 1992), suggesting that sensitization depends on the presence of a low concentration of capsaicin near the terminals of the nociceptors. The primary heat hyperalgesia is presumably due to peripheral sensitization and it lasts for 1 to 3 hours (LaMotte, et al., 1991).

An expanding area of secondary mechanical hyperalgesia and mechanical allodynia develops concentrically around the injection site and zone of primary hyperalgesia during a period of about 15-60 minutes after damage to the skin (Figure 6.1; Lewis, 1942; Hardy, et al., 1952; LaMotte, et al., 1991; 1992). Following an intradermal injection of 100 μg of capsaicin, the zones of secondary hyperalgesia and allodynia reach their maximum extents in about 15-30 minutes.

The maximum area of the zone of secondary hyperalgesia is greater than that of the area of allodynia, suggesting different underlying mechanisms. The allodynia then recedes slowly until it disappears after about 2 hours; the secondary hyperalgesia may last up to 24 hours (Simone, et al., 1989; LaMotte, et al., 1991;Torebjörk, et al., 1992). The secondary mechanical hyperalgesia can best be demonstrated using punctate mechanical stimulation with a von Frey filament about 1-2 cm from the injection site (LaMotte, et al., 1991). Mechanical allodynia can be shown by the presence of tenderness in response to gentle strokes of the skin applied radially at 2 cm from the injection site.

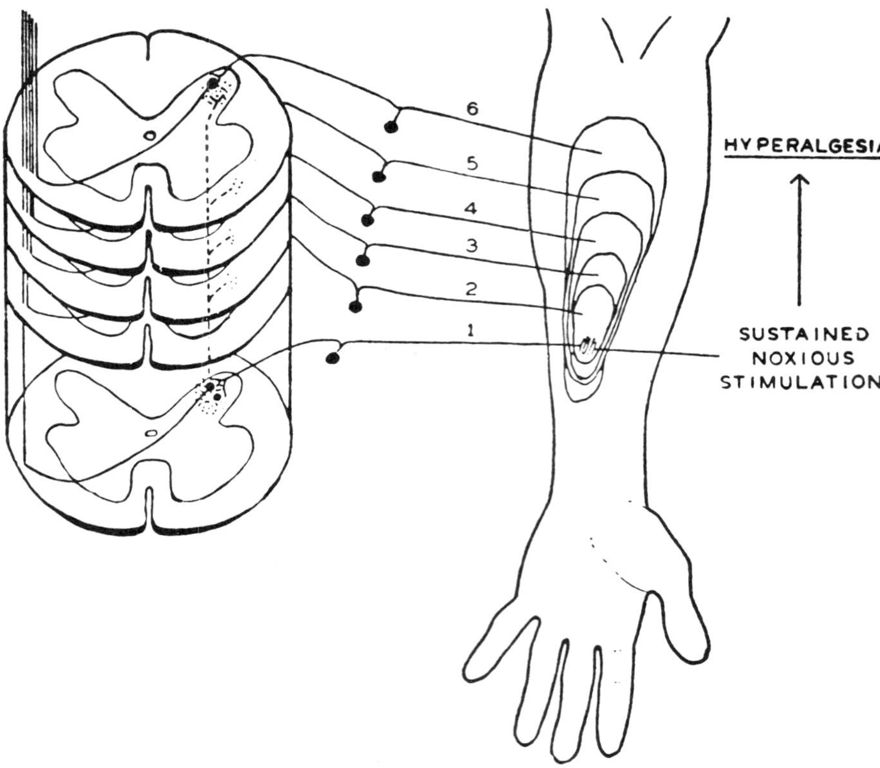

Figure 6.1. Development of secondary hyperalgesia and allodynia after a noxious stimulus. Application of a strong noxious stimulus to the skin at the site indicated by the hatched region on the drawing of the forearm at the right results in the progressive concentric spread of hyperalgesia and allodynia over a region of skin surrounding the area of damage. Primary afferent nociceptors that supply the damaged area of skin would activate nociceptive dorsal horn neurons, such as spinothalsamic tract (STT) cells, in the appropriate segment of the spinal cord, as shown in the drawing at the left. However, central processing through interneuronal circuits also leads to an enhanced responsiveness of STT cells in adjacent parts of the dorsal horn (indicated by the stippled areas in the spinal cord drawing) that now become more responsive to stimulation of undamaged skin. The altered responses reflect central sensitization, and the increased input to higher centers in the brain would account for secondary hyperalgesia and allodynia. (From Hardy, et al., 1967.)

The initiation of secondary mechanical hyperalgesia and allodynia depends on the activation of C fibers (Torebjörk, et al., 1992). By contrast, the expression of mechanical allodynia depends on the activation of large, myelinated mechanoreceptor afferents and of secondary mechanical hyperalgesia on stimulation of mechanically activated nociceptors (Koltzenburg, et al., 1992; Treede and Cole, 1993; Cervero, et al., 1994).

The term "secondary" is used because secondary hyperalgesia and allodynia are not due to a direct action of capsaicin on peripheral nerve fibers but rather are caused by central nervous system mechanisms. This is demonstrated in part by the fact that the sensitivity of the afferent fibers that supply the skin in the secondary zone is not altered by the capsaicin injection (Figure 6.2; Baumann, et al., 1991; LaMotte, et al., 1992). Curiously, secondary heat hyperalgesia does not occur following an intradermal injection of capsaicin (Ali, et al., 1996), despite the fact that central nociceptive neurons are sensitized. However, consistent with the psychophysical observations, the responses of central neurons to noxious heat are also not enhanced (see below).

In addition to these sensory changes, an area of vasodilation or "flare" is observed following the intradermal injection of capsaicin (LaMotte, et al., 1991; cf., Lewis, 1927). However, there is no wheal. After an injection of 100 μg of capsaicin into the skin, flare reaches its maximum extent within 10 minutes and disappears in 1-2 hours (LaMotte, et al., 1991). The changes in responses to mechanical stimulation of the skin in the area of secondary hyperalgesia and allodynia are independent of the flare (Simone, et al., 1989; Ali, et al., 1996; Andrews, et al., 1999).

6.3 CENTRAL SENSITIZATION: COMPARISONS WITH WINDUP, CENTRAL SUMMATION AND LONG-TERM POTENTIATION

Recordings of the enhanced responses of a central nociceptive neuron to a constant input from peripheral stimulation can demonstrate central sensitization after a strong or prolonged noxious stimulus. However, central sensitization needs to be distinguished from windup and central summation. It may involve mechanisms similar to those of long-term potentiation.

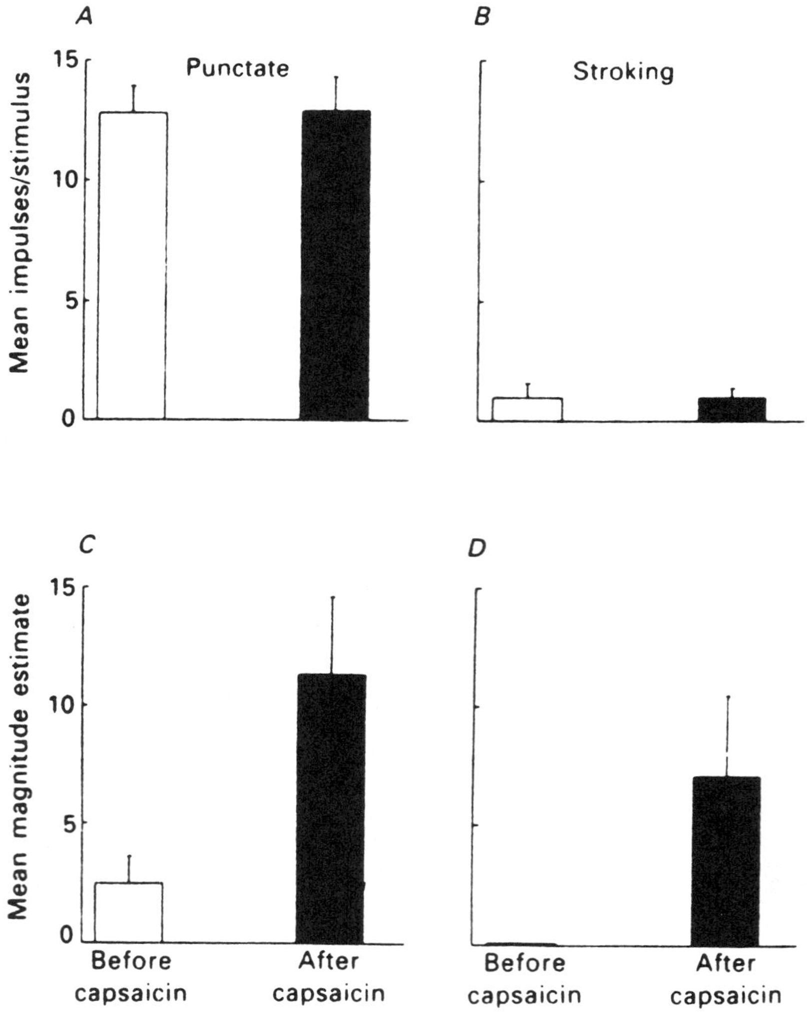

Figure 6.2. The effects of an intradermal injection of capsaicin in human subjects on the mean responses of C-nociceptors and on pain responses to punctate stimulation and to stroking of the skin in the area of secondary hyperalgesia and allodynia in the same subjects. Capsaicin was injected 3-30 mm outside of the receptive fields of C-afferent nociceptors. Receptive field size (not illustrated) and the responses to standard noxious ansd innocuous mechanical stimuli (A, B) were unaltered by the capsaicin injection. However, the magnitude estimates of pain evoked by the same stimuli showed large increases (C, D). (From LaMotte, et al., 1992.)

Windup

Low frequency repetitive stimulation of afferent C fibers can result in the "windup" of responses of dorsal horn nociceptive neurons (Mendell, 1966), including STT cells (Chung, et al., 1979). It has been shown that windup depends upon the temporal summation of long-lasting excitatory postsynaptic potentials (EPSPs) that are produced by the actions of excitatory amino acids and substance P on non-NMDA, NMDA and neurokinin receptors (Davies and Lodge, 1987; Dickenson and Sullivan, 1987; Thompson, et al., 1992; 1993; Nagy, et al, 1994). C fiber volleys at frequencies as low as once per 2 to 3 seconds can produce windup (Mendell, 1966). This is accounted for by the fact that the EPSPs evoked by stimulation of C fibers last for seconds. However, in addition to temporal summation of postsynaptic depolarizations, windup leads to a short-term potentiation that persists for a few minutes after stimulation of C fibers is terminated and after the membrane potential resumes its original level (Jeftinija and Urban, 1994). This is a homosynaptic type of potentiation, since it does not affect EPSPs evoked by C-fibers in dorsal roots that are adjacent to the one that was stimulated to produce the windup. This suggests that windup involves mechanisms in the part of the pathway that is presynaptic to the recorded neuron (Jeftinija and Urban, 1994). Windup thus differs from central sensitization in several respects: it lasts a shorter length of time and it is not heterosynaptic. Windup can trigger central sensitization, but central sensitization does not require prior windup (Woolf, 1996). Instead, central sensitization depends on an increase in intracellular calcium, which can be produced by a variety of mechanisms that do not necessarily require action potential production (Woolf, 1996).

Central Summation

If a noxious stimulus sensitizes nociceptive afferent fibers, these fibers may become spontaneously active and thus provide a continuous input to the central nervous system that may affect the excitability of the central nociceptive neuron under observation. For instance, this continuous nociceptive input might produce a maintained increase in the excitability of the central neuron by depolarizing it, causing it to have enhanced responses to a variety of inputs from other afferent fibers. An example of this would be the responses of dorsal horn interneurons to stimulation of the skin in the peripheral part of the receptive field (Woolf and King, 1989). Intracellular recordings show that excitatory responses to mechanical stimulation in the central part of the receptive field or "firing zone" evoke excitatory postsynaptic potentials (EPSPs) that exceed threshold for action potential generation, as well as subthreshold EPSPs. However, stimulation of the skin outside the "firing zone", a region Woolf and King term the "low-probability firing fringe," results in EPSPs that are usually subthreshold but that occasionally fire action potentials when an EPSP sums with a membrane depolarization from some other source. Thus, the ability of an EPSP that is evoked from the "low-probability firing fringe" of the receptive field to cause a discharge depends on summation with a depolarization that is produced by another input. This type of enhanced responsiveness of central nociceptive neurons does not represent central sensitization but rather a form of spatial summation. Temporal summation could also occur, although other processes intervene when C fibers are activated repetitively (see discussion of windup and post-tetanic potentiation).

At least two lines of evidence suggest that STT neurons develop central sensitization, rather than just showing central summation, following an intradermal injection of capsaicin. First, capsaicin causes a transient activation of capsaicin-sensitive primary afferent nociceptors that lasts only 3 to 10 minutes in monkeys, following which the afferents become desensitized and cease discharging (Baumann, et al, 1991; LaMotte, et al., 1992). This means that an intradermal injection of capsaicin would not result in a continuous input to the spinal cord dorsal horn. This contrasts with the effects of an injection of formalin into the skin, which produces a long-lasting discharge of nociceptive afferent fibers (Puig and Sorkin, 1996). Secondly, STT neurons do not develop an enhanced responsiveness to all types of inputs during central sensitization, in contrast to what might be expected during central summation. They do have increased responses to innocuous and weakly noxious mechanical stimuli applied to an area of skin well away from the injection site, the presumed area of secondary mechanical hyperalgesia and allodynia (Simone, et al., 1991; Dougherty and Willis, 1992; Lin, et al., 1996a; Sluka, et al., 1997a). However, they generally do not show enhanced responses to intense noxious mechanical stimuli (although some individual STT cells do have increases in their responses to strong noxious mechanical stimuli; Dougherty, et al., 1992; Lin, et al., 1999a; see Figures 6.3 and 6.4). Their responses to noxious heat stimuli applied in the same region are actually reduced (Sluka, et al., 1997a). Following capsaicin injection, STT cells also respond with an increased number of discharges following stimulation of a dorsal rootlet (Simone, et al., 1991) and during iontophoretic applications of excitatory amino acids (Dougherty and Willis, 1992). However, they show reduced responses to iontophoretic release of inhibitory amino acids (Lin, et al., 1996a). Inhibition would still be expected if central summation were occurring.

Long-term Potentiation

Central sensitization of the responses of STT cells to innocuous and weakly noxious stimuli following an intradermal injection of capsaicin, as well as secondary mechanical allodynia, lasts only about 1.5 to 2 hours (LaMotte, et al., 1991; Dougherty, et al., 1992). However, secondary mechanical hyperalgesia can last up to 24 hours (LaMotte, et al., 1991). Electrical stimulation of C fibers at a high frequency can evoke long lasting increases in C fiber-evoked field potentials in the dorsal horn (Liu and Sandkühler, 1997). For example, stimulation of the sciatic nerve in rats at 100 Hz, using a strength that activates both A and C fibers (400 pulses given in 4 trains of 1 second duration at 10 second intervals) results in an enhancement in C fiber-evoked field potentials that lasts for 4-9 hours. The time course of this potentiation closely resembles that of long-term potentiation in other parts of the central nervous system. Stimulation of C fibers at a lower frequency (20 or 2 Hz) results in a potentiation of the field potentials that lasts 3-6 hours. There is no potentiation when the stimulus frequency is only 1 Hz (although windup of action potentials in dorsal horn neurons can be observed in the same preparations when this frequency of stimulation is used). The duration of the central sensitization that follows intradermal capsaicin injection may relate to the frequency and duration of the C fiber discharges produced by the action of capsaicin on particular types of primary afferent nociceptors that contain VR1 receptors. Another, perhaps more important, factor is that the descending control systems have been left intact in the experiments on central sensitization of STT cells. Stimulation of C fibers by natural forms of noxious stimuli can evoke long-term potentiation only if the spinal cord is transected. With the spinal cord intact, noxious stimuli do not evoke long-term potentiation, indicating that descending inhibitory systems interfere with the expression of long-term potentiation in the spinal cord (Sandkühler and Liu, 1998).

6.4 CENTRAL SENSITIZATION OF STT CELLS

Examples of the responses of STT neurons to mechanical stimulation of the skin before and after an intradermal injection of capsaicin are shown in Figures 6.3 and 6.4 (Dougherty and Willis, 1992). The neuron illustrated in Figures 6.3 was classified as a wide dynamic range (WDR) STT cell (Mendell, 1966), since it responded to both innocuous (brush) and noxious (pressure and pinch) mechanical stimuli. Initially, the neuron had very little background activity (Figure 6.3A), and it responded only weakly to brush and pressure stimuli applied to several sites on the skin (Figure 6.3C and E). There was a stronger response to noxious pinch (Figure 6.3G). The injection of capsaicin resulted in an increase in background activity (Figure 6.3B) and enhanced responses to all of the mechanical stimuli (Figure 6.3D, F and H). The receptive field also expanded (Figure 6.3, drawing at bottom).

The STT neuron whose responses are shown in Figure 6.4 was classified as a high threshold (HT) cell, since it was not activated by innocuous mechanical stimuli (Chung, et al., 1979; 1986). In fact, brush stimuli produced an inhibition of the background activity (Figure 6.4C), and so this neuron could be classified as a high-threshold inhibitory STT cell, according to the classification scheme suggested by Blair, et al. (1981). Capsaicin injection resulted in a transient increase followed by a transient decrease in the background activity of the cell (Figure 6.4B). The responses to pressure stimuli were increased (cf., Figure 6.4E and F), but the responses to brushing and pinch were not substantially altered (Figure 6.4C and D, G and H). The size of the receptive field increased (Figure 6.4, drawing at bottom).

In the summary bar graph in Figure 6.5A (Dougherty, et al., 1992), the mean background activity and the responses of a sample population of 25 spinothalamic tract cells to brush (innocuous) and pressure (weakly noxious) stimuli were increased, but not the responses to noxious pinch. The sample was then subdivided into WDR cells and HT cells. Interestingly, the responses of 18 WDR spinothalamic tract cells to innocuous and weakly noxious mechanical stimuli were significantly increased (Figure 6.5B), whereas the responses of 7 HT spinothalamic tract cells to the same stimuli were not significantly changed (Figure 6.5C), although there was a trend for the responses to pressure to increase. Evidently,

WDR STT neurons are more likely to demonstrate central sensitization than are HT STT cells. This observation suggests that HT STT neurons are less likely to be involved in plastic changes than are WDR STT cells. As a consequence, WDR cells seem more likely to contribute to secondary hyperalgesia and allodynia than are HT STT cells. Perhaps HT STT cells are chiefly responsible for signaling acute pain and WDR cells play a more important role in altered pain states.

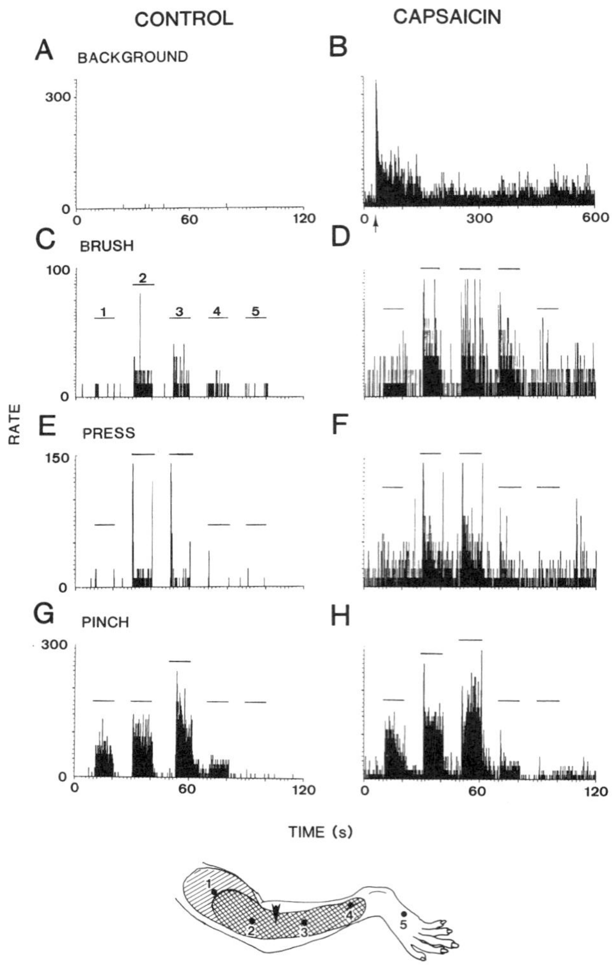

Figure 6.3. Changes in the activity of a wide dynamic range STT neuron following an intradermal injection of capsaicin. In A is shown the background activity of the cell and in B the response to the capsaicin injection. Responses to brush, press and pinch stimuli applied to points 1-5 on the skin of the lower extremity before capsaicin injection are shown in C, E and G and after capsaicin injection in D, F and H. The drawing at the bottom shows the receptive field before (doubly hatched area) and after (total hatched area) capsaicin injection. The numbers indicate the points at which the mechanical stimuli were applied. The arrowhead shows the injection site. (From Dougherty and Willis, 1992.)

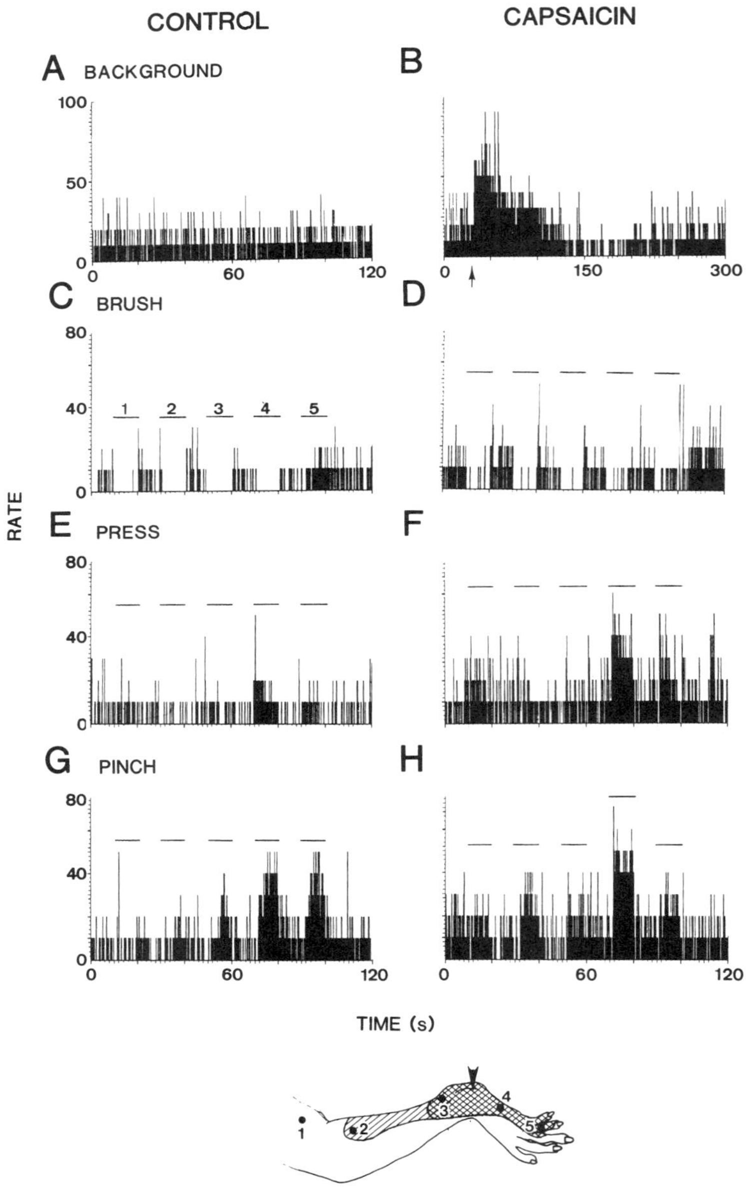

Figure 6.4. Changes in the activity of a high threshold STT neuron following an intradermal injection of capsaicin. In A is shown the background activity of the cell and in B the response to the capsaicin injection. C, E and G are the responses to brush, press and pinch stimuli applied to points 1-5 on the skin of the lower extremity before the capsaicin injection, and D, F and H are the responses after the injection. The drawing at the bottom shows the receptive field before (doubly hatched area) and after (total hatched area) capsaicin injection. The numbers indicate the points stimulated, and the arrowhead indicates the injection site. (From Dougherty and Willis, 1992.)

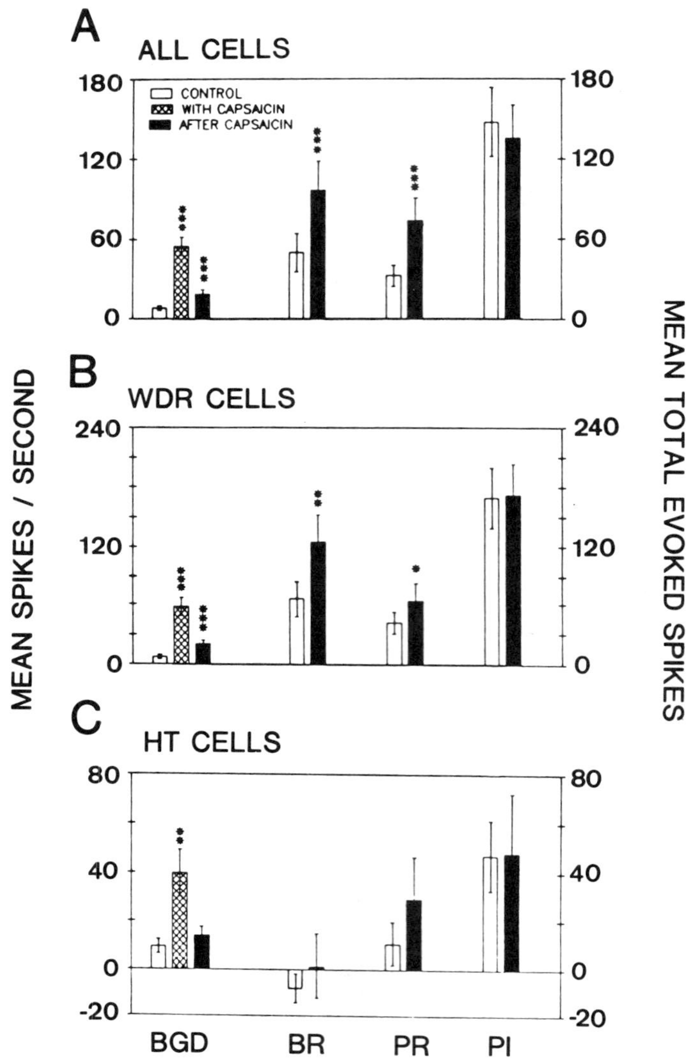

Figure 6.5. Summary of the background activity (BGD) and responses of a population of STT cells to brush (BR), press (PR) and pinch (PI) stimuli before and after an intradermal injection of capsaicin. In A are shown the responses of all of the cells in the sample. In B and C are the responses of wide dynamic range (WDR) and high threshold (HT) STT cells. The asterisks indicate statistically significant changes. (From Dougherty and Willis, 1992.)

6.5 NEUROTRANSMITTERS AND RECEPTORS THAT TRIGGER CENTRAL SENSITIZATION FOLLOWING INTRADERMAL CAPSAICIN INJECTION

A clue as to the neurotransmitters that are likely to be released in the spinal cord following an intradermal injection of capsaicin comes from recent studies of the capsaicin or VR1 receptor. A large fraction of the primary afferent fibers that contain VR1 (capsaicin) receptors also contain substance P (Tominaga, et al., 1998). Furthermore, glutamate is co-localized in the terminals in the spinal cord dorsal horn of primary afferent fibers that contain substance P (De Biasi and Rustioni, 1988). Thus, it is not surprising that application of capsaicin to spinal cord slices results in the release of substance P (Gamse, et al., 1979) and that intradermal injection of capsaicin results in the release of glutamate in the dorsal horn of the spinal cord (Sorkin and McAdoo, 1993). Aspartate is also released, perhaps from dorsal horn interneurons whose activity is increased following capsaicin injection. Calcitonin gene-related peptide (CGRP) is likely to be released centrally from primary afferent terminals, since CGRP is released from the peripheral terminals of afferent fibers by capsaicin (see review by Willis, 1999). There is also evidence for the release of neurokinin A (NKA) in the dorsal horn following capsaicin activation of primary afferent neurons (Nagy, et al., 1993).

Since capsaicin-sensitive afferent fibers release glutamate and substance P in the spinal cord, these substances are obvious candidates for the trigger that initiates central sensitization. Receptors that are potentially involved in the actions of glutamate (and aspartate) include AMPA, kainate and NMDA ionotropic glutamate receptors and metabotropic glutamate receptors (Conn, et al., 1994; Dingledine, et al., 1999). The action of substance P is mediated largely by NK_1 receptors and of NKA by NK_2 receptors (Nakanishi, 1991; Nagy, et al., 1993; Krause, et al., 1994). Recently it has been found that spinal cord NK_3 receptors can contribute to thermal hyperalgesia (but not mechanical hyperalgesia) in the rat spinal cord (Linden and Seybold, 1999). The role of NK_3 receptors, if any, in central sensitization of STT cells following an intradermal capsaicin injection has not yet been determined.

The excitatory amino acid agonist, NMDA, can act cooperatively with substance P to produce a long-lasting increase in the responses of STT neurons to NMDA or to mechanical stimulation of the cutaneous receptive field (Dougherty and Willis, 1991). Repeated iontophoretic applications of NMDA result in progressively decreasing responses of STT cells (Figure 6.6G), and iontophoretic release of substance P alone has only a minimal excitatory action. However, when substance P is released for 30 seconds, the later responses of some STT cells to NMDA can increase progressively for about 15-20 minutes (Figure 6.6C and D) and then remain increased for hours (Figure 6.6E and F; Dougherty and Willis, 1991). When tested during the period of enhanced NMDA responses, the responses of the same STT neurons to innocuous and noxious stimulation of the cutaneous receptive field are also increased (Dougherty and Willis, 1991). The responses of STT neurons to AMPA can also increase following a substance P application, but for only about 15-30 minutes (Dougherty, et al., 1993). Not all STT neurons show increases in responses to excitatory amino acids following co-administration with substance P. Sometimes the responses to excitatory amino acids decrease (Dougherty, et al., 1993).

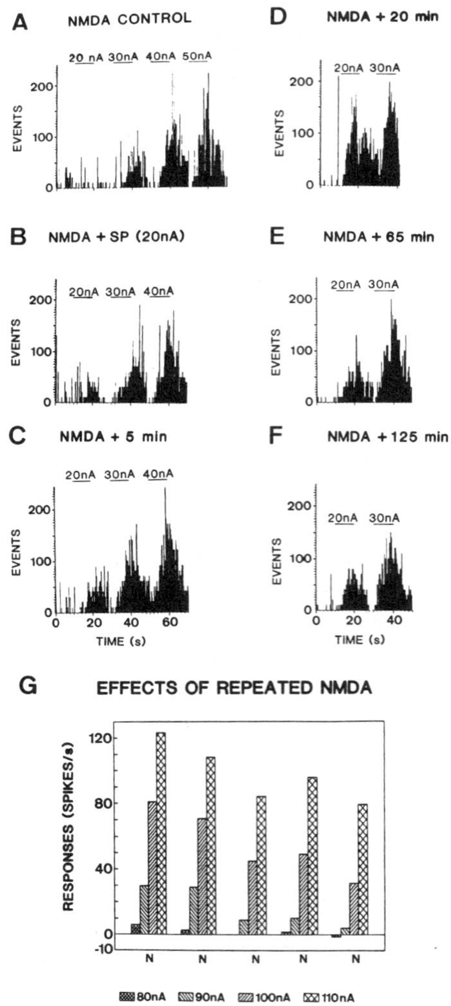

Figure 6.6. Cooperative action of NMDA and substance P (SP) in sensitizing an STT cell. In A-F are shown the responses of an STT cell to graded iontophoretic doses of NMDA. The responses in A are controls. The responses in B are those recorded during the continuous release of SP. C-F show the responses to NMDA at various times after the SP application was discontinued. In G are the effects of repeated applications of graded doses of NMDA in the absence of SP. (From Dougherty and Willis, 1991.)

The main experimental approach used to examine the role of excitatory amino acid and substance P receptors in the central sensitization of STT cells was administration of antagonists of non-NMDA, NMDA, and metabotropic glutamate receptors and of NK_1 receptors into the dorsal horn by microdialysis. One advantage of administering drugs by microdialysis is that the drug actions are confined to a restricted volume of the spinal cord (Sluka and Westlund, 1993) in the vicinity of the recorded STT neuron. Intrathecal drug administration would affect a much larger extent of the spinal cord. Furthermore, drugs released by microdialysis can diffuse throughout much of the dorsal horn, thereby affecting much of the dendritic tree of the neuron under investigation (as well as local interneurons and synaptic terminals). Drugs released by iontophoresis affect only structures within a radius of about 200 μm from the tip of the electrode (Herz, et al., 1969).

Microdialysis administration of the non-NMDA glutamate receptor antagonist, CNQX, largely blocked the responses of STT cells to intradermal injection of capsaicin and to mechanical stimulation of the skin. Presumably because of this, CNQX prevented the development or expression of central sensitization (Figure 6.7, left sets of bars in top row of panels; Dougherty, et al., 1992). However, the NMDA glutamate receptor antagonist, AP7, had no effect on the responses of STT neurons to innocuous and weakly noxious mechanical stimuli and yet prevented the development of central sensitization of the responses to these stimuli (Figure 6.7, right sets of bars in top row of panels).

Similarly, administration of the NK_1 receptor antagonists, CP96345 or GR82334, had no effect on the responses of STT neurons to mechanical stimulation of the skin but prevented the central sensitization of these responses by capsaicin injection (Figure 6.7, second row of panels; Dougherty, et al., 1994). CP96344, which has similar actions to CP96345 except that it does not antagonize NK_1 receptors, had no effect (Figure 6.7, right sets of bars, lower row of panels). The NK_2 antagonists, MEN10376 and GR94800, did not appear to prevent central sensitization (not illustrated), and so NK_2 receptors do not appear to be involved.

In recent experiments, the effects of AIDA and CPCCOEt, which are antagonists of group I metabotropic glutamate receptors, were observed to reduce the responses of spinothalamic tract cells to noxious, but not to innocuous, mechanical stimulation of the skin (Figure 6.8A; Neugebauer, et al., 1999). During the central sensitization produced by intradermal injection of capsaicin, the responses to innocuous and noxious mechanical stimuli were reduced transiently by brief microdialysis administration of these drugs (Figure 6.8B).

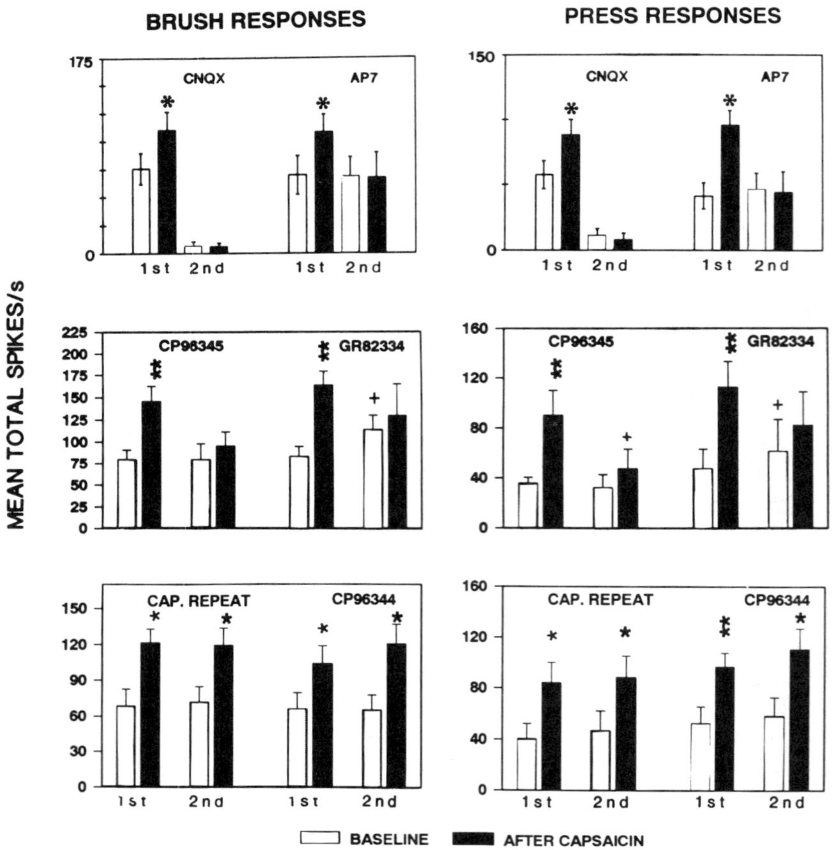

Figure 6.7. Effects of glutamate and neurokinin receptor antagonists on central sensitization following intradermal injection of capsaicin. The bars represent the mean summed responses of STT cells to brush and press stimuli applied to the receptive fields. Open bars are control responses and filled bars are the responses after capsaicin. The left pair of bars in each panel represent responses during microdialysis of artificial cerebrospinal fluid into the dorsal horn. The right pair of bars in the top 2 rows represent responses during microdialysis of the antagonist indicated, including the non-NMDA glutamate receptor antagonist, CNQX, the NMDA receptor antagonist, AP7, and the NK_1 receptor antagonists, CP96345 and GR 82334. The responses in the lower row represent those observed after repeated injections of capsaicin (CAP repeat) or during the microdialysis administration of CP 96344 (inactive isomer of CP 96345). (From Dougherty, et al., 1992; 1994.)

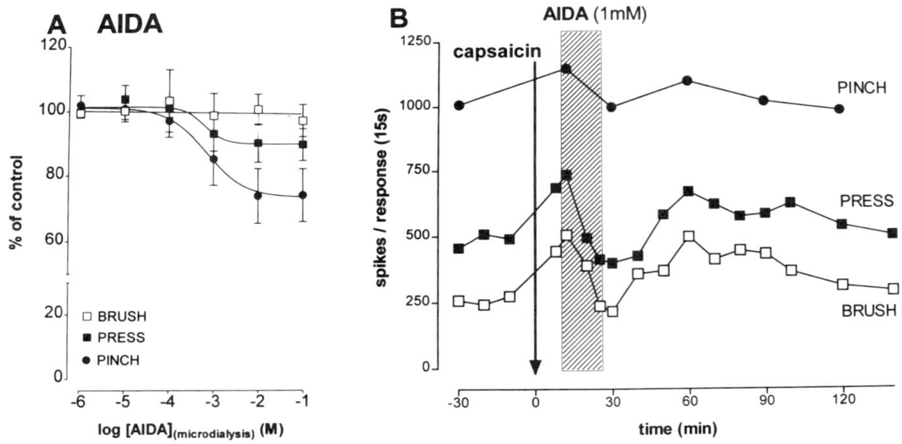

Figure 6.8. Effects of a group I metabotropic glutamate receptor antagonist on the responses of STT cells to mechanical stimulation. In A are shown the dose-dependent actions of AIDA, an mGluR1 antagonist, on the mean responses of 6 WDR STT neurons to brush, press and pinch stimuli. In B, the responses of an individual STT cell are shown. The responses to brush, press and pinch stiuli were enhanced following an intradermal injection of capsaicin. Microdialysis administration of AIDA for 15 minutes transiently reduced the central sensitization of all 3 responses. After the drug was washed out, the responses increased again. (From Neugebauer, et al., 1999.)

In summary, non-NMDA, NMDA and metabotropic glutamate receptors and NK_1 receptors all play a role in triggering central sensitization of primate spinothalamic tract cells following intradermal injection of capsaicin. However, this does not rule out a possible contribution by other substances acting at their receptors, such as CGRP acting at CGRP receptors.

6.6 ROLE OF SIGNAL TRANSDUCTION CASCADES IN CENTRAL SENSITIZATION

We believe that central sensitization of STT cells depends on the activation of signal transduction pathways that secondarily alter the responsiveness of STT cells (in agreement with Woolf, 1996). For this reason, we have investigated the effects of administration of a number of agents that affect signal transduction pathways to determine which pathways play a role in central sensitization.

Protein Kinase C Pathway

The first signal transduction cascade that we examined was the protein kinase C pathway. This seemed a reasonable point of departure, since several other laboratories have demonstrated a role in nociception of this pathway (Coderre, 1992; Mao, et al., 1993; Malmberg, et al.,1997b).

The most direct evidence that protein kinase C is involved in central sensitization is that the increases in the responses of STT cells to mechanical stimulation of the skin following intradermal injection of capsaicin are blocked by microdialysis administration of a selective protein kinase C inhibitor, NPC15437 (Figure 6.9A; Lin, et al., 1996c; Sluka, et al., 1997a, b). Furthermore, central sensitization can be mimicked by microdialysis administration of an active phorbol ester (Figure 6.9B), but not by an inactive one (Figure 6.9C), into the dorsal horn (Palecek, et al., 1994; Lin, et al., 1996c). Infusion of an active phorbol ester (but not an inactive phorbol ester) into the dorsal horn by microdialysis not only causes increased responses of STT cells to innocuous and weakly noxious mechanical stimuli but also increases in the concentrations of glutamate and aspartate (as well as of glycine and taurine) in the dorsal horn (Palecek, et al., 1999).

These observations suggest that the activation of protein kinase C may help mediate secondary mechanical hyperalgesia and allodynia.

Nitric Oxide/Protein Kinase G Pathway

Another signal transduction cascade that appears to play a role in the central sensitization of spinothalamic tract cells after intradermal injection of capsaicin is the nitric oxide/protein kinase G pathway. Nitric oxide has been implicated in pain mechanisms in a number of studies (Kitto, et al., 1992; Meller, et al., 1992; 1994; Moore, et al., 1991). The target of nitric oxide in neural tissue is often guanylyl cylcase (Bredt and Snyder, 1989), which is an enzyme that increases intracellular concentrations of cGMP. The cGMP in turn activates protein kinase G (Bredt and Synder, 1989; see reviews by Meller and Gebhart, 1993; Lincoln, et al., 1994).

The sensitization of spinothalamic tract cells after intradermal injection of capsaicin can be blocked by nitric oxide synthase inhibitors, such as L-NAME (Figure 6.10A; Lin, et al., 1999a), or by the guanylyl cyclase inhibitor, ODQ (Figure 6.10B; Lin, et al., 1997a). Furthermore, the increase in the concentrations of the excitatory amino acids, glutamate and aspartate, in the dorsal horn that is produced by intradermal injection of capsaicin can be blocked by microdialysis administration of the protein kinase G inhibitor, KT5823 (Sluka and Willis, 1998).

Capsaicin injection causes the release of nitric oxide in the dorsal horn, as shown by increased concentrations of its metabolites, NO_2^- and NO_3^-, in dialysate that has passed through a microdialysis fiber inserted across the dorsal horn (Wu, et al., 1998; Lin, et al., 1999a). The increase in nitric oxide concentration lasts for about 2 hours. The effects of capsaicin injection in producing central sensitization can be mimicked by microdialysis administration of SIN-1, a nitric oxide donor, or of 8-bromo-cGMP (Lin, et al., 1999a). The action of SIN-1 is mediated through an action on guanylyl cyclase, since the action of SIN-1 is blocked by the guanylyl cyclase inhibitor, ODQ (Lin, et al., 1999c). The responses to both weak and strong noxious mechanical stimuli, but not to noxious heat, are increased. This suggests that the NO/cGMP cascade may play a role in secondary mechanical hyperalgesia and allodynia.

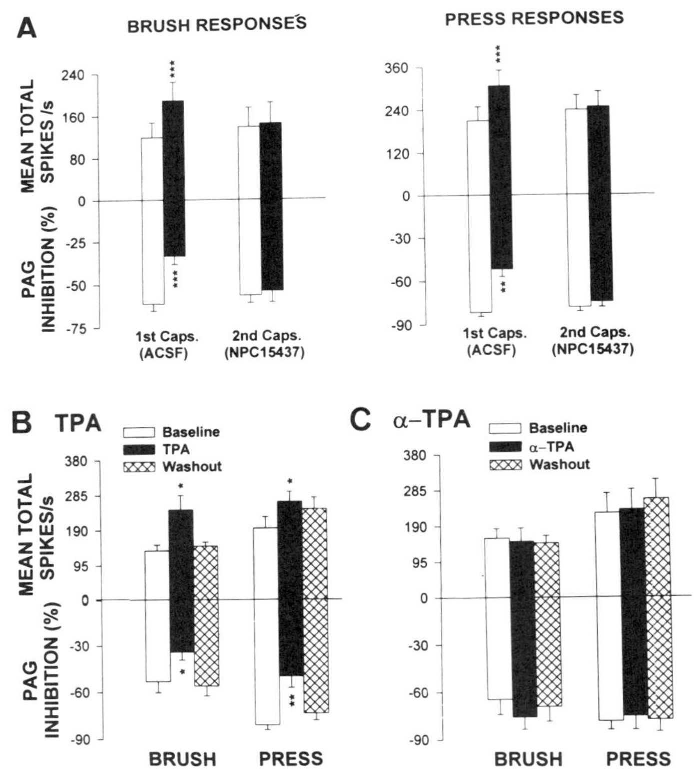

Figure 6.9. Comparison of central sensitization produced by intradermal injection of capsaicin and the effects of microdialysis administration of active and inactive phorbol esters. The upgoing bars show the mean responses of a sample of STT neurons to brush and press stimuli and the downgoing bars the percent inhibition of these responses produced by stimulation in the periaqaueductal gray (PAG). In A are shown the effects of two capsaicin injections (Caps.). Artificial cerebrospinal fluid (ACSF) was infused into the dorsal horn by microdialysis during the first capsaicin injection and the PKC inhibitor, NPC15437, during the second injection. In B, the effects of TPA (active phorbol ester) infusion are shown. In C, the lack of effcct of α-TPA (inactive phorbol ester) is shown. (From Lin, et al., 1997c.)

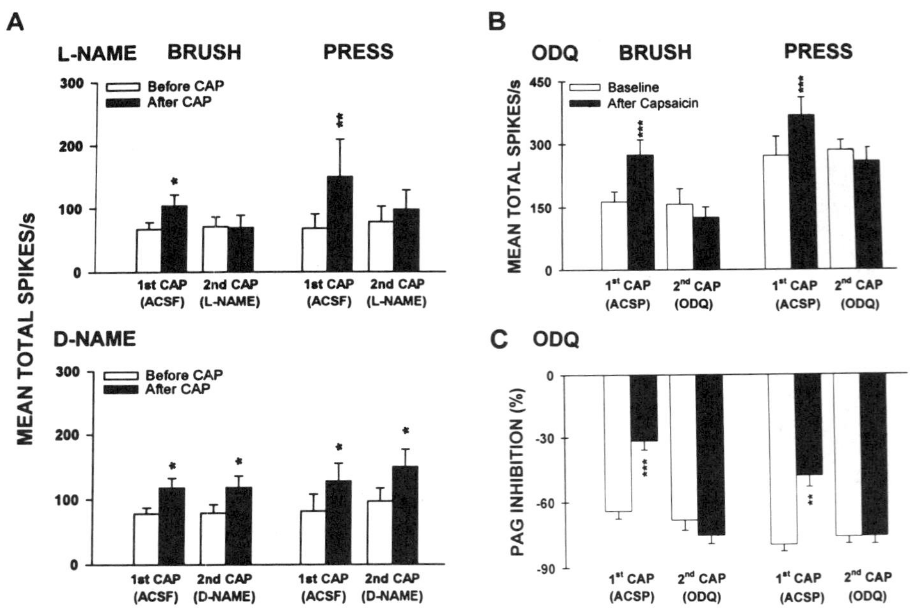

Figure 6.10. The role of nitric oxide and guanylyl cyclase in central sensitization. The pairs of bars indicate the mean responses of a sample of STT cells to brush and press stimuli before and after each of two injections of capsaicin (CAP). In the upper graph of A are shown the effect of the nitric oxide synthase inhibitor, L-NAME, in blocking the enhancement of responses following the second capsaicin injection. The lower graph shows that the inactive isomer, D-NAME, has no effect. The upper graph in B shows that the guanylyl cyclase inhibitor, ODQ, blocks the enhanced responses to the second capsaicin injection, and the lower graph shows that ODQ prevents much of the inhibition of the responses produced by stimulation in the periaqueductal gray. (From Lin, et al., 1997a; 1999a.)

Protein Kinase A Pathway

Another signal transduction pathway that contributes to nociceptive processing is the protein kinase A pathway. Central sensitization of STT neurons following intradermal capsaicin injection can be blocked by microdialysis administration into the dorsal horn of the selective PKA inhibitor, H89 (Sluka, et al., 1997a). Central sensitization can be in part mimicked by administration of forskolin, and the effects of forskolin are blocked by H89 (Lin, et al., 1997b). However, instead of enhancing chiefly the responses to innocuous and weakly noxious mechanical stimuli, forskolin increases the responses to noxious mechanical stimuli, suggesting that the protein kinase A pathway is more likely to be involved in producing secondary mechanical hyperalgesia than allodynia (Lin, et al., 1997b). Furthermore, forskolin administration did not block the inhibition of STT cells that is produced by stimulation in the periaqueductal gray (Lin, et al., 1997b), and forskolin was not consistently effective in preventing the inhibition produced by iontophoretic release of GABA or glycine (Lin, et al., 1998).

6.7 ACTIONS OF PROTEIN KINASE INHIBITORS ON BEHAVIORAL RESPONSES OF RATS FOLLOWING INTRADERMAL CAPSAICIN

Injection of capsaicin into the plantar skin of the foot in rats failed to produce any signs of secondary heat hyperalgesia, as judged by the paw withdrawal latency to noxious heat stimuli (Sluka and Willis, 1997). However, there was a marked increase in the rate at which the rats withdrew from innocuous mechanical stimuli (von Frey filaments applied to the foot outside the injected area), suggesting that the rats develop secondary mechanical allodynia after capsaicin injection (Figure 6.11A). The allodynia was transiently blocked during microdialysis administration of a general protein kinase inhibitor (H7), a protein kinase C inhibitor (Figure 6.11B; NPC15437), a protein kinase A inhibitor (Figure 6.11C; H89) or a protein kinase G inhibitor (Figure 6.11D; KT5823).

These results suggest that in rats 1) changes that take place in nociceptive processing within the spinal cord following intradermal capsaicin injection result in secondary mechanical allodynia; 2) secondary heat hyperalgesia does not develop following intradermal capsaicin injection, 3) protein kinases C, G and A all contribute to secondary mechanical allodynia, and 4) the allodynia is maintained by a dynamic process that can be temporarily interrupted by posttreatment with protein kinase inhibitors.

6.8 MECHANISM OF INCREASED RESPONSIVENESS OF SPINOTHALAMIC TRACT NEURONS DURING CENTRAL SENSITIZATION

It is unclear what is the site of action for drugs administered into the dorsal horn by microdialysis. The action could be on receptors located on presynaptic endings, on interneurons or on the STT neurons themselves. Localization is somewhat better defined for the actions of drugs released by iontophoresis, although there are the same target options.

Nevertheless, two major alterations in the responses of STT neurons to iontophoretically released excitatory and inhibitory amino acids can be detected during central sensitization. There are 1) an increase in the responses to excitatory amino acids (Figure 6.12A; Dougherty and Willis, 1992) and 2) a decrease in the responses to inhibitory amino acids (Figure 6.12B; Lin, et al., 1996a; 1999b). In parallel with the reduction in responses to inhibitory amino acids, there is a decrease in the ability of stimulation in the periaqueductal gray to inhibit the sensitized neurons (Figures 6.9 and 6.10; Lin, et al., 1996a; 1999b). Thus, STT neurons should have increased responses to excitatory amino acids that are released by the activation of primary afferent fibers, either from afferent terminals or from interneurons, and they should have reduced responses to inhibitory actions mediated by pathways descending from the brain or by local inhibitory circuits. It is unclear at present if inhibitory actions mediated by neurotransmitters other than inhibitory amino acids are affected during central sensitization. For example, the inhibition evoked by stimulation in the periaqueductal gray is thought to involve the release of serotonin and norepinephrine in the dorsal horn, as well as of inhibitory amino acids (Peng, et al., 1996a, b, c; Lin, et al., 1996; Cui, et al., 1999).

A possible mechanism by which the responses to excitatory amino acids could become enhanced and cause central sensitization would be through phosphorylation of the receptors or their ion channels. Chen and Huang (1991) have shown that intracellular injection of protein kinase C increases the responsiveness of isolated neurons of the trigeminal nucleus caudalis to NMDA. They proposed that phosphorylation of the NMDA receptors causes an allosteric change in the conformation of the ion channel associated with the NMDA receptor and that this change results in a reduced binding of Mg^{++} (Chen and Huang, 1992). The result is an increase in the responsiveness of NMDA receptors. Conversely, phosphorylation of inhibitory amino acid receptors can inhibit their chloride currents (cf., Leidenheimer, et al., 1992; Vaello, et al., 1994). The result would be a reduced responsiveness of these receptors to the action of inhibitory amino acids. As already mentioned, it is possible that other inhibitory receptors may also be involved.

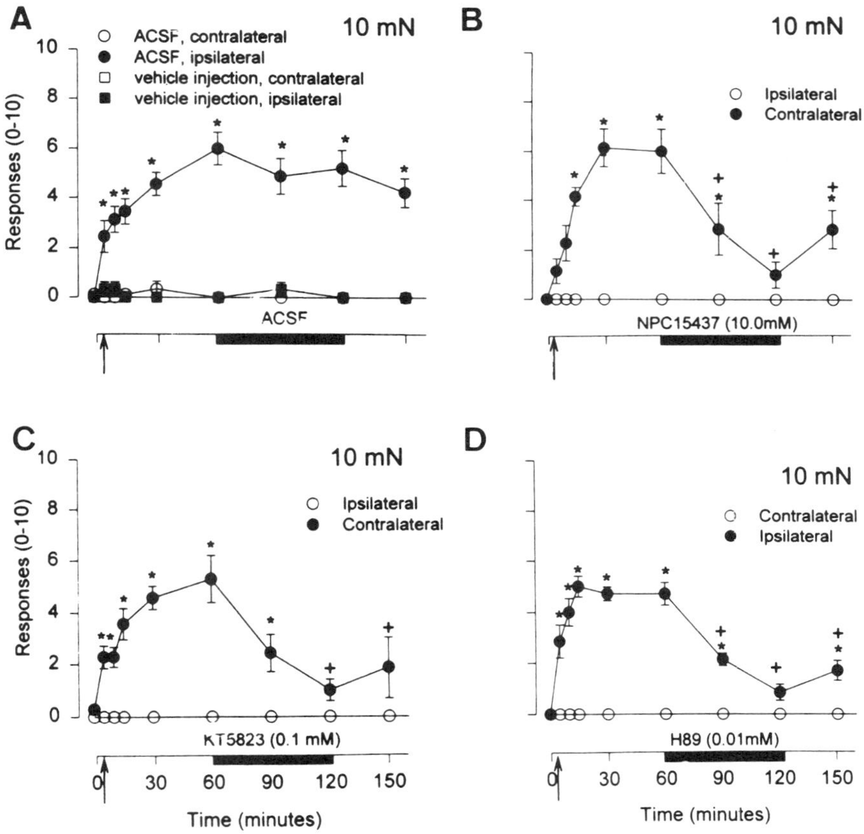

Figure 6.11. Action of protein kinase inhibitors on the mechanical allodynia produced by intradermal injection of capsaicin in rats. The graphs show the proportion of paw withdrawal responses to 10 applications of a von Frey filament with a 10 mN bending force to the plantar surface of the foot ipsilateral or contralateral to an injection of capsaicin. In A, ACSF was infused into the dorsal horn during the time indicated by the horizontal bar. In B, the dialysate contained an inhibitor of protein kinase C (NPC15437); in C, a protein kinase G inhibitor was given (KT5823); and in D a protein kinase A inhibitor (H89) was used. (From Sluka and Willis, 1997.)

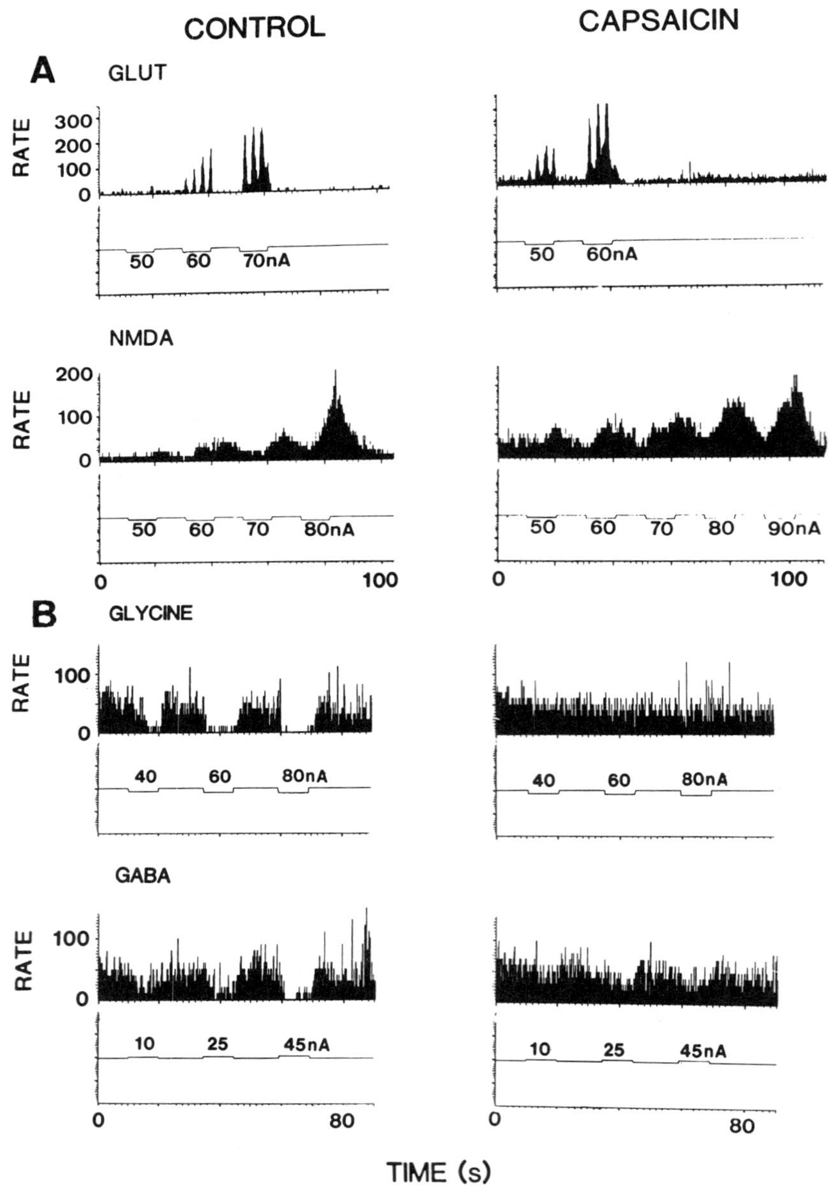

Figure 6.12. Effect of intradermal injection of capsaicin on the responses of STT neurons to iontophoretically released excitatory and inhibitory amino acids. The peristimulus time histograms show the graded effects of different iontophoretic current doses of amino acids released in the vicinity of STT cells before and after an intradermal injection of capsaicin. In A are shown increases in the excitatory effects of glutamate (GLUT) and NMDA, whereas in B are shown the inhibitory effects of glycine and GABA. (From Dougherty and Willis, 1992; Lin, et al, 1996a.)

An observation that may be pertinent to the parallel increases in excitatory amino acid responses and reductions in inhibitory amino acid responses is the action of group I metabotropic glutamate receptors in enhancing AMPA receptor activity and reducing $GABA_A$ responses evoked by muscimol in the nucleus of the solitary tract (Glaum and Miller, 1993). It was suggested that this action might be mediated by the protein kinase G pathway, since a similar effect was produced by administration of 8-bromo-cGMP.

6.8 POSSIBILITY OF LONG-TERM CHANGES

The responses of STT neurons in monkeys to innocuous and weakly noxious stimuli are enhanced for only about 1.5 to 2 hours after intradermal injection of capsaicin. This duration is similar to that of secondary mechanical allodynia in human subjects (LaMotte, et al., 1991), somewhat less than that of the enhanced behavioral responses of rats to mechanical stimulation of the foot (Sluka and Willis, 1997), and considerably less than the duration of secondary mechanical hyperalgesia in humans (up to 24 hours). Nevertheless, it seems possible that events involving changes in gene expression are set in motion by intradermal injection of capsaicin that could have longer term implications (cf., Ji and Rupp, 1997; Herdegan and Leah, 1998). Evidence for this is that an intradermal capsaicin injection causes increased expression of c-fos in the dorsal horn of the spinal cord, and this is blocked by the nitric synthase inhibitor, L-NAME (Wu, et al., 2000). Furthermore, microdialysis adminstration of SIN-1 increases c-fos in the dorsal horn, which can be blocked by L-NAME (Wu, et al., 2000), as well as phosphorylation of cAMP-responsive element-binding protein (CREB) in the dorsal horn (Wu, et al., 1999). Presumably, these actions are mediated by the release of nitric oxide in the spinal cord.

Although it is unclear that the activation of constitutive and inducible transcription factors (Herdegan and Leah, 1998) results in any significant changes in nociceptive processing in the dorsal horn following injection of capsaicin, there is clearly potential for changes in gene expression related to synaptic structure, the numbers and subunit composition of neurotransmitter receptors and voltage gated ion channels, as well as activation, inhibition or production of other proteins. Presumably, a transient change, such as that due to capsaicin injection, might become a permanent change following a prolonged noxious stimulus.

6.9 IS CENTRAL SENSITIZATION A TYPE OF LONG-TERM POTENTIATION?

Long-term potentiation is defined as an enduring increase in synaptic efficacy. It is generally evoked by high frequency electrical stimulation, but it has also been produced by other means. The duration of the central sensitization of responses of primate STT cells to innocuous and weakly noxious mechanical stimuli and of human secondary mechanical allodynia following intradermal injection of capsaicin is 1-2 hours, that of rat secondary mechanical allodynia more than 3 hours and that of human secondary mechanical hyperalgesia up to 24 hours. Thus, central sensitization has a prolonged time course, like long-term potentiation (see Riedel, et al., 1996). Central sensitization and long-term potentiation have in common a role of glutamate acting at non-NMDA, NMDA and metabotropic glutamate receptors (see above; McEachern and Shaw, 1996). A key factor in both is an increase in intracellular Ca^{++} concentration. Furthermore, central sensitization and long-term potentiation both involve the activation of the protein kinase C, NO/protein kinase G and protein kinase A signal transduction pathways (see above; McEachern and Shaw, 1996). Long-term potentiation also involves other signal transduction pathways that have not yet been shown to have a role in central sensitization but that may well be found to contribute. Both central sensitization and long-term potentiation may be associated with changes in gene expression that can produce very long lasting alterations in synaptic transmission (see above; McEachern and Shaw, 1996).

For these reasons, central sensitization resembles long-term potentiation in many respects.

6.10 CONCLUSIONS

1. The model chosen for studying central sensitization was intradermal injection of capsaicin. Advantages of this model are: the ability to correlate neural responses with human psychophysics; the possibility of comparing the changes in the responses of central nociceptive neurons to one of two injections of capsaicin that would otherwise have comparable effects; and the absence of continual afferent input after the initiating event, making it easier to study central sensitization in the absence of central summation.

2. An intradermal injection of capsaicin in human subjects produces: pain; an anesthetic bleb; a surrounding area of primary hyperalgeia; expanding and then contracting concentric zones of secondary mechanical hyperalgesia and allodynia; and an area of flare. Interestingly, secondary heat hyperalgesia does not occur. Secondary hyperalgesia and allodynia depend on changes in the responsiveness of central nociceptive neurons, a process termed central sensitization. These events are initiated by a burst of activity in C fibers. However, secondary hyperalgesia depends on activation of mechanical nociceptors and allodynia on activation of mechanorceptors supplied by large myelinated axons.

3. Spinothalamic tract (STT) neurons develop increased background activity and enhanced responses to innocuous and weakly noxious mechanical stimuli applied to the presumed area of secondary hyperalgesia and allodynia following intradermal injection of capsaicin. Responses to

noxious heat stimuli are reduced. Wide dynamic range STT neurons are more likely to become sensitized than are high threshold STT cells.

4. The effects of the administration of receptor antagonists by microdialysis suggest a role for glutamate (and perhaps aspartate) and neurokinins, such as substance P, in triggering central sensitization. The receptors involved include non-NMDA, NMDA and metabotropic glutamate receptors and neurokinin receptors. Other substances, such as calcitonin-gene-related peptide, and their receptors could also be involved.

5. Several different signal transduction cascades have been shown to contribute to the maintenance of central sensitization following an intradermal injection of capsaicin. These include the protein kinase C, NO/protein kinase G, and protein kinase A pathways.

6. Behavioral signs of secondary mechanical allodynia can be demonstrated in rats following intradermal injection of capsaicin. This allodynia can be blocked by inhibitors of protein kinase C, protein kinase G and protein kinase A.

7. The enhanced responses of STT cells during central sensitization following capsaicin injection are associated with increased responses to iontophoretically released excitatory amino acids and decreased responses to iontophoretically released inhibitory amino acids. Both changes would be expected to contribute to the enhanced responses. It is suggested that these changes reflect phosphorylation of the receptors or their associated channels.

8. Long term changes could also be initiated following capsaicin injection by activation of constitutive or inducible transcription factors and consequent alterations in gene expression. Although the alterations in spinal cord activity that are produced by capsaicin injection appear to be short-lived, more prolonged noxious stimulation could lead to permanent changes that affect the excitability of nociceptive neurons and hence pain processing.

9. There are a number of parallels between central sensitization in spinal cord nociceptive circuits and long-term potentiation, including the time course, initiating events, maintenance by signal transduction pathways and changes in gene expression.

6.11 REFERENCES

Ali, Z., Meyer, R.A.. and Campbell, J.N. Secondary hyperalgesia to mechanical but not heat stimuli following a capsaicin injection in hairy skin. *Pain,* 1996; 68:401-411.

Andrews, K., Baranowski, A.. and Kinnman, E. Sensory threshold changes without initial pain or alterations in cutaneous blood flow, in the area of secondary hyperalgesia caused by topical application of capsaicin in humans. *Neuroscience Letters* 266:45-48, 1999.

Baumann, T.K., Simone, D.A., Shain, C.N. and LaMotte, R.H. Neurogenic hyperalgesia: The search for the primary cutaneous afferent fibers that contribute to capsaicin-induced pain and hyperalgesia. *Journal of Neurophysiology*, 1991; 66:212-227.

Blair, R.W., Weber, R.N. and Foreman, R.D. Characteristics of primate spinothalamic tract neurons receiving viscerosomatic convergent input in T_3-T_5 segments. *Journal of Neurophysiology,* 1981; 46:797-811.

Bredt, D.S. and Snyder, S.H. Nitric oxide mediates glutamate-linked enhancement of cGMP levels in the cerebellum. *Proceedings of The National Academy of Sciences of the United States of America*, 1989; 86:9030-9033.

Carpenter, S.E. and Lynn, B. Vascular and sensory responses of human skin to mild injury after topical treatment with capsaicin. *British Journal of Pharmacology*,1981; 73:755-758.

Cervero, F., Meyer, R.A. and Campbell, J.N. A psychophysical study of secondary hyperalgesia: evidence for increased pain to input from nociceptors. *Pain,* 1994; 58:21-28.

Chen, L. and Huang, L.Y.M. Sustained potentiation of NMDA receptor-mediated glutamate responses through activation of protein kinase C by a μ opioid. *Neuron,* 1991; 7:319-326.

Chen, L. and Huang, L.Y.M. Protein kinase C reduced Mg^{++} block of NMDA-receptor channels as a mechanism of modulation. *Nature,* 1992; 356:521-523.

Chung, J.M., Kenshalo, D.R., Gerhart, K.D. and Willis, W.D. Excitation of primate spinothalamic neurons by cutaneous C-fiber volleys. *Journal of Neurophysiology,* 1979; 42:1354-1369.

Chung, J.M., Surmeier, D.J., Lee, K.H., Sorkin, L.S., Honda, C.N., Tsong, Y. and Willis, W.D. Classification of primate spinothalamic and somatosensory thalamic neurons based on cluster analysis. *Journal of Neurophysiology,* 1986; 56:308-327.

Coderre, T.J. Contribution of protein kinase C to central sensitization and persistent pain following tissue injury. *Neuroscience Letters,* 1992; 140:181-184.

Conn, P.J., Boss, V. and Chung, D.S. (1994). Second-messenger systems coupled to metabotropic glutamate receptors. In: P.J. Conn and J. Patel (Eds.), *The Metabotropic Glutamate Receptors* (pp. 59-98). Totowa, N.J.: Human Press Inc.

Cui, M., Feng, Y., McAdoo, D.J. and Willis, W.D. Periaqueductal gray stimulation-induced inhibition of nociceptive dorsal horn neurons in rats is associated with the release of norepinephrine, serotonin, and amino acids. *Journal of Pharmacology and Experimental Therapeutics,* 1999; 289:868-876.

Daniel, H., Levenes, C. and Crepel, F. Cellular mechanisms of cerebellar LTD. *Trends in Neuroscience,* 1998; 21:401-407.

Davies, S.N. and Lodge, D. Evidence for involvement of N-methylaspartate receptors in 'wind-up' of class 2 neurones in the dorsal horn of the rat. *Brain Research,* 1987; 424:402-406.

De Biasi, S. and Rustioni, A. Glutamate and substance P coexist in primary afferent terminals in the superficial laminae of spinal cord. *Proceedings of The National Academy of Sciences of the United States of America*, 1988; 85:7820-7824.

Dickenson, A.H. and Sullivan, A.F. Evidence for a role of the NMDA receptor in the frequency dependent potentiation of deep rat dorsal horn nociceptive neurones following C fibre stimulation. *Neuropharmacology,* 1987; 26:1235-1238.

Dingledine, R., Borges, K., Bowie, D. and Traynelis, S.F. The glutamate receptor ion channels. *Pharmacology Review* 51:7-61, 1999.

Dougherty, P.M. and Willis, W.D. Enhancement of spinothalamic neuron responses to chemical and mechanical stimuli following combined micro-iontophoretic application of N-methyl-D-aspartic acid and substance P. *Pain,* 1991; 47:85-93.

Dougherty P.M. and Willis, W.D. Enhanced responses of spinothalamic tract neurons to excitatory amino acids accompany capsaicin-induced sensitization in the monkey. *Journal of Neuroscience,* 1992; 12:883-894.

Dougherty, P.M., Palecek, J., Paleckova, V., Sorkin, L.S. and Willis, W.D. The role of NMDA and non-NMDA excitatory amino acid receptors in the excitation of primate spinothalamic tract neurons by mechanical, chemical, thermal, and electrical stimuli. *Journal of Neuroscience,* 1992; 12:3025-3041.

Dougherty, P.M., Palecek, J., Paleckova, V. and Willis, W.D. Neurokinin 1 and 2 antagonists attenuate the responses and NK1 antagonists prevent the sensitization of primate spinothalamic tract neurons after intradermal capsaicin. *Journal of Neurophysiology,* 1994; 72:1464-1475.

Dougherty, P.M., Palecek, J., Zorn, S. and Willis, W.D. Combined application of excitatory amino acids and substance P produces long-lasting changes in responses of primate spinothalamic tract neurons. *Brain Research Review,* 1993; 18:227-246.

Dubner R. and Ruda, M.A.. Activity-dependent neuronal plasticity following tissue injury and inflammation. *Trends in Neuroscience,* 1992; 15:96-103.

Gamse, R., Molnar, A. and Lembeck, F. Substance P release from spinal cord slices by capsaicin. *Life Sciences,* 1979; 25:629-636.

Glaum, S.R. and Miller, R.J. Activation of metabotropic glutamate receptors produces reciprocal regulation of ionotropic glutamate and GABA responses in the nucleus of the tractus-solitarius of the rat. *Journal of Neuroscience,* 1993; 13:1636-1641.

Hardy, J.D., Wolff, H.G. and Goodell, H. *Pain sensations and reactions.* The Williams & Wilkins Co., N.Y., 1952; reprinted by Hafner Publishing Company, Inc., New York, 1967.

Herdegen, T. and Leah, J.D. Inducible and constitutive transcription factors in the mammalian nervous system: control of gene expression by Jun, Fos and Krox, and CREB/ATF proteins. *Brain Research Review,* 1998; 28:370-490.

Herz, A., Zieglgänsberger, W. and Färber, G. Microelectrophoretic studies concerning the spread of glutamic acid and GABA in brain tissue. *Experimental Brain Research,* 1969; 9:221-235.

Hölscher, C. Long-term potentiation: a good model for learning and memory? *Progress In Neuro-Psychopharmacology and Biological Psychiatry,* 1997; 21:47-68.

Jeftinija, S. and Urban, L. Repetitive stimulation induced potentiation of excitatory transmission in the rat dorsal horn: an in vitro study. *Journal of Neurophysiology,* 1994; 71:216-228.

Ji, R.R. and Rupp, F. Phosphorylation of transcription factor CREB in rat spinal cord after formalin-induced hyperalgesia: relationship to c-fos induction. *Journal of Neuroscience,* 1997; 17:1776-1785.

Kenshalo, D.R., Leonard, R.B., Chung, J.M. and Willis, W.D. Facilitation of the responses of primate spinothalamic cells to cold and to tactile stimuli by noxious heating of the skin. *Pain,* 1982; 12:141-152.

Kitto, K.F., Haley, J.E. and Wilcox, G.L. Involvement of nitric oxide in spinally mediated hyperalgesia in the mouse. *Neuroscience Letters,* 1992; 148:1-5.

Koltzenburg, M., Lundberg, L.E.R. and Torebjörk, H.E. Dynamic and static components of mechanical hyperalgesia in human hairy skin. *Pain,* 1992; 51:207-219.

Krause, J.E., Sachais, B.S. and Blount, P. Tachykinin. (1994). Receptors. In Peroutka, S.J (Ed.), *Handbook of Receptors and Channels; G Protein Coupled Receptors.* Boca Raton: CRC Press.

LaMotte, R.H., Lundberg, L.E.R. and Torebjörk, H.E. Pain, hyperalgesia and activity in nociceptive C units in humans after intradermal injection of capsaicin. *Journal of Physiology,* 1992; 448:749-764.

LaMotte, R.H., Shain, C.H., Simone, D.A. and Tsai, E.F.P. Neurogenic hyperalgesia: Psychophysical studies of underlying mechanisms. *Journal of Neurophysiology,* 1991; 66:190-211.

Leidenheimer, N.J., McQuilkin, S.J., Hahner, L.D., Whiting, P. and Harris, R.A. Activation of protein-kinase-C selectively inhibits the gamma-aminobutyric $acid_A$ receptor: role of desensitization. *Molecular Pharmacology,* 1992; 41:1116-1123.

Lewis, T. *The blood vessels of the human skin and their responses.* Shaw & Sons Ltd., London, 1927.

Lewis, T. *Pain.* The Macmillan Press Ltd, London, 1942.

Lin, Q., Palecek, J., Paleckova, V., Peng, Y.B., Wu, J., Cui, M. and Willis, W.D. Nitric oxide mediates the central sensitization of primate spinothalamic tract neurons. *Journal of Neurophysiology,* 1999; 81:1075-1085.

Lin, Q., Peng, Y.B. and Willis, W.D. Inhibition of primate spinothalamic tract neurons by spinal glycine and GABA is reduced during central sensitization. *Journal of Neurophysiology,* 1996a; 76:1005-1014.

Lin, Q., Peng, Y.B. and Willis, W.D. Antinociception and inhibition from the periaqueductal gray are mediated in part by spinal 5-$hydroxytryptamine_{1A}$ receptors. *Journal of Pharmacology and Experimental Therapeutics,* 1996b; 276:958-967.

Lin, Q., Peng, Y.B. and Willis, W.D. Possible role of protein kinase C in the sensitiization of primate spinothalamic tract neurons. *Journal of Neuroscience,* 1996c; 16:3026-3034.

Lin, Q., Peng, Y.B., Wu, J. and Willis, W.D. Involvement of cGMP in nociceptive processing by and sensitization of spinothalamic neurons in primates. *Journal of Neuroscience,* 1997; 17:3293-3302.

Lin, Q., Wu, J., Cui, M. and Willis, W.D. Protein kinase A sensitizes primate spinothalamic tract neurons by influencing spinal excitatory and inhibitory amino acid receptors. *Neuroscience Abstracts,* 1998; 24:637.

Lin, Q., Wu, J., Peng, Y.B., Cui, M. and Willis, W.D. Nitric oxide-mediated spinal disinhibition contributes to the sensitization of primate spinothalamic tract neurons. *Journal of Neurophysiology*, 1999a; 81:1086-1094.

Lin, Q., Wu, J., Peng, Y.B., Cui, M. and Willis, W.D. Inhibition of primate spinothalamic tract neurons by spinal glycine and GABA is modulated by guanosine 3', 5'-cyclic monophosphate. *Journal of Neurophysiology,* 1999b; 81:1095-1103.

Lin, Q., Wu, J. and Willis, W.D. The effect of protein kinase A activation on the responses of primate spinothalamic neurons to mechanical and thermal stimuli. *Neuroscience Abstracts,* 1997; 23:2357.

Lincoln, T.M., Komalavilas, P. and Cornwell, T.L. Pleiotropic regulation of vascular smooth muscle tone by cyclic GMP-dependent protein kinase. *Hypertension,* 1994; 23:1141-1147.

Linden, D.R. and Seybold, V.S. Spinal neurokinin$_3$ receptors mediate thermal but not mechanical hyperalgesia via nitric oxide. *Pain,* 1999; 80:309-317.

Liu, X. and Sandkühler, J. Characterization of long-term potentiation of C-fiber-evoked potentials in spinal cord dorsal horn of adult rat: essential role of NK1 and NK2 receptors. *Journal of Neurophysiology,* 1997; 78:1973-1982.

Malmberg, A.B., Brandon, E.P., Idzerda, R.L., Liu, H., McKnight, G.S. and Basbaum, A.I. Diminished inflammation and nociceptive pain with preservation of neuropathic pain in mice with a targeted mutation of the type I regulatory subunit of cAMP-dependent protein kinase. *Journal of Neuroscience,* 1997; 17:7462-7470.

Malmberg, A.B., Chen, C., Tonegawa, S. and Basbaum, A.I. Preserved acute pain and reduced neuropathic pain in mice lacking PKCgamma. *Science,* 1997; 278:279-283.

Mao, J., Mayer, D.J., Hayes, R.L. and Price, D.D. Spatial patterns of increased spinal cord membrane-bound protein kinase C and their relation to increases in ^{14}C-2-deoxyglucose metabolic activity in rats with painful peripheral mononeuropathy. *Journal of Neurophysiology,* 1993; 70:470-481.

McEachern, J.C. and Shaw, C.A. An alternative to the LTP orthodoxy: a plasticity-pathology continuum model. *Brain Research Review,* 1996; 22:51-92.

Meller, S.T. and Gebhart, G.F. Nitric oxide (NO) and nociceptive processing in the spinal cord. *Pain,* 1993; 52:127-136.

Meller, S.T., Cummings, C.P., Traub, R.J. and Gebhart, G.F. The role of nitric oxide in the development and maintenance of the hyperalgesia produced by intraplantar injection of carrageenan in the rat. *Neuroscience,* 1994; 60:367-374.

Meller, S.T., Pechman, P.S., Gebhart, G.F. and Maves, T.J. Nitric oxide mediates the thermal hyperalgesia produced in a model of neuropathic pain in the rat. *Neuroscience,* 1992; 50:7-10.

Mendell, L.M. Physiological properties of unmyelinated fiber projection to the spinal cord. *Experimental Neurology*, 1966; 16:316-332.

Merskey, H. and Bogduk, N. (Eds.) *Classification of chronic pain. Descriptions of chronic pain syndromes and definitions of pain terms.* 2nd ed. IASP Press, Seattle, 1994.

Meyer, R.A. and Campbell, J.N. Myelinated nociceptive afferents account for the hyperalgesia that follows a burn to the hand. *Science,* 1981; 213:1527-1529.

Moore, P.K., Babbedge, R.C., Wallace, P., Gaffen, Z.A. and Hart, S.L. 7-nitro indazole, an inhibitor of nitric oxide synthase, inhibits anti-nociceptive activity in the mouse without increasing blood pressure. *British Journal of Pharmacology*, 1993; 108:296-297.

Nagy, I., Maggi, C.A., Dray, A., Woolf, C.J. and Urban, L. The role of neurokinin and N-methyl-D-aspartate receptors in synaptic transmission from capsaicin sensitive primary afferents in the rat spinal cord *in vitro*. *Neuroscience,* 1993; 52:1029-1037.

Nagy, I., Miller, B.A. and Woolf, C.J. NK_1 and NK_2 receptors contribute to C-fibre evoked slow potentials in the rat spinal cord. *Neuroreport,* 1994; 5:2105-2108.

Nakanishi, S. Mammalian tachykinin receptors. *Annual Review of Neuroscience*, 1991; 14:123-136.

Neugebauer, V., Chen, P.S. and Willis, W.D. Role of metabotropic glutamate receptor subtype mGluR1 in brief nociception and central sensitization of primate STT cells. *Journal of Neurophysiol*ogy, 1999; 82: 272-282.

Palecek, J., Paleckova, V., Dougherty, P.M. and Willis, W.D. The effect of phorbol esters on the responses of primate spinothalamic neurons to mechanical and thermal stimuli. *Journal of Neurophysiol*ogy, 1994; 71:529-537.

Palecek, J., Paleckova, V. and Willis, W.D. The effect of phorbol esters on spinal cord amino acid concentrations and responsiveness of rats to mechanical and thermal stimuli. *Pain,* 1999; 80:597-605.

Peng, Y.B., Lin, Q. and Willis, W.D. The role of $5\text{-}HT_3$ receptors in periaqueductal gray-induced inhibition of nociceptive dorsal horn neurons in rats. *Journal of Pharmacology and Experimental Therapeutics,* 1996a; 276:116-124.

Peng, Y.B., Lin, Q. and Willis, W.D. Involvement of *alpha*-2 adrenoreceptors in the periaqueductal gray-induced inhibition of dorsal horn cell activity in rats. *Journal of Pharmacology and Experimental Therapeutics,* 1996b; 278:125-135.

Peng, Y.B., Lin, Q. and Willis, W.D. Effects of GABA and glycine receptor antagonists on the activity and PAG-induced inhibition of rat dorsal horn neurons. *Brain Research*, 1996c; 736:189-201.

Puig, S. and Sorkin, L.S. Formalin-evoked activity in identified primary afferent fibers: systemic lidocaine suppresses phase-2 activity. *Pain,* 1996; 64:345-355.

Raja, S.N., Meyer, R.A. and Campbell, J.N. Peripheral mechanisms of somatic pain. *Anesthesiology,* 1988; 68:571-590.

Randic, M., Hecimovic, H. and Ryu, P.D. Substance P modulates glutamate-induced currents in acutely isolated rat spinal dorsal horn neurones. *Neuroscience Letters*, 1990; 117:74-80.

Randic, M., Jiang, M.C. and Cerne, R. Long-term potentiation and long-term depression of primary afferent neurotransmission in the rat spinal cord. *Journal of Neuroscience,* 1993; 13:5228-524.

Riedel, G., Wetzel, W. and Reymann, K.G. Comparing the role of metabotropic glutamate receptors in long-term potentiation and in learning and memory. *Progress in Neuro-Psychopharmacology and Biological Psychiatry,* 1996; 20:761-789.

Sandkühler, J. and Liu, X. Induction of long-term potentiation at spinal synapses by noxious stimulation or nerve injury. *European Journal of Neuroscience,* 1998; 10:2476-2480.

Simone, D.A., Baumann, T.K. and LaMotte, R.H. Dose-dependent pain and mechanical hyperalgesia in humans after intradermal injection of capsaicin. *Pain,* 1989; 38:99-107.

Simone, D.A., Ngeow, J.Y.F., Putterman, G.J. and LaMotte, R.H. Hyperalgesia to heat after intradermal injection of capsaicin. *Brain Research,* 1987; 418:201-203.

Simone, D.A., Nolano, M., Johnson, T., Wendelschafer-Crabb, G. and Kennedy, W.R. Intradermal injection of capsaicin in humans produces degeneration and subsequent reinnervation of epidermal nerve fibers: correlation with sensory function. *Journal of Neuroscience,* 1998; 18:8947-8959.

Simone, D.A., Sorkin, L.S., Oh, U., Chung, J.M., Owens, C., LaMotte, R.H. and Willis, W.D. Neurogenic hyperalgesia: central neural correlates in responses of spinothalamic tract neurons. *Journal of Neurophysiology* ,61991; 6:228-246.

Sluka, K.A. and Westlund, K.N. Centrally administered non-NMDA but not NMDA receptor antagonists block peripheral knee joint inflammation. *Pain,* 1993; 55:217-225.

Sluka, K.A.. and Willis, W.D. The effects of G-protein and protein kinase inhibitors on the behavioral responses of rats to intradermal injection of capsaicin. *Pain,* 1997; 71:165-178.

Sluka, K.A.. and Willis, W.D. Increased spinal release of excitatory amino acids following intradermal injection of capsaicin is reduced by a protein kinase G inhibitor. *Brain Research,* 1998; 798:281-286.

Sluka, K.A.., Rees, H.., Chen, P.S., Tsuruoka, M. and Willis, W.D. Inhibitors of G-proteins and protein kinases reduce the sensitization to mechanical stimulation and the desensitization to heat of spinothalamic tract neurons induced by intradermal injection of capsaicin in the primate. *Experimental Brain Research,* 1997a; 115:15-24.

Sluka, K.A., Rees, H., Chen, P.S., Tsuruoka, M. and Willis, W.D. Capsaicin-induced sensitization of primate spinothalamic tract cells is prevented by a protein kinase C inhibitor. *Brain Research,* 1997b; 772:82-86.

Sorkin, L.S. and McAdoo, D.J. Amino acids and serotonin are released into the lumbar spinal cord of the anesthetized cat following intradermal capsaicin injections. *Brain Research,* 1993; 607:89-98.

Svendsen, F., Tjølsen, A. and Hole, K. LTP of spinal Aβ and C-fibre evoked responses after electrical sciatic nerve stimulation. *Neuroreport, 1997;* 8:3427-3430.

Thompson, R.F., Bao, S., Chen, L., Cipriano, B.D., Grethe, J.S., Kim, J.J., Thompson, J.K., Tracy, J.A., Weninger, M.S. and Krupa, D.J. Associative learning. *International Review of Neurobiology,* 1997; 41:151-189.

Thompson, S.W.N., Gerber, G., Sivilotti, L.G. and Woolf, C.J. Long duration ventral root potentials in the neonatal rat spinal cord in vitro; the effects of ionotropic and metabotropic excitatory amino acid receptor antagonists. *Brain Research,* 1992; 595:87-97.

Thompson, S.W.N., Urban, L. and Dray, A. Contribution of NK_1 and NK_2 receptor activation to high threshold afferent fibre evoked ventral root responses in the rat spinal cord in vitro. *Brain Research,* 1993; 625:100-108.

Tominaga, M., Caterina, M.J., Malmberg, A.B., Rosen, T.A., Gilbert, H., Skinner, K., Raumann, B.E., Basbaum, A.I. and Julius, D. The cloned capsaicin receptor integrates multiple pain-producing stimuli. *Neuron,* 1998; 21:531-543.

Torebjörk, H.E., Lundberg, L.E.R. and LaMotte, R.H. Central changes in processing of mechanoreceptive input in capsaicin-induced secondary hyperalgesia in humans. *Journal of Physiology,* 1992; 448:765-780.

Treede, R.D. and Cole, J.D. Dissociated secondary hyperalgesia in a subject with large-fibre sensory neuropathy. *Pain,* 1993; 53:169-174.

Vaello, M.L., Ruiz-Gómez, A., Lerma, J. and Mayor, F. Modulation of inhibitory glycine receptors by phosphorylation by protein kinase C and cAMP-dependent protein kinase. *Journal of Biological Chemistry,* 1994; 269:2002-2008.

Willis, W.D. Mechanical allodynia. A role for sensitized nociceptive tract cells with convergent input from mechanoreceptors and nociceptors? *APS Journal,* 1993; 2:23-33.

Willis, W.D. Is central sensitization of nociceptive transmission in the spinal cord a variety of long-term potentiation? A commentary on the article by Svendsen, Tjølsen and Hole. In Focus, *Neuroreport,* 1997; 8:iii.

Willis, W.D. Dorsal root potentials and dorsal root reflexes: a double-edged sword. *Experimental Brain Research,* 1999; 124:395-421.

Windhorst, U. The spinal cord and its brain: representations and models. To what extent do forebrain mechanisms appear at brainstem and spinal cord levels? *Progress In Neurobiology,* 1996; 49:381-414.

Woolf, C.J. Evidence for a central component of post-injury pain hypersensitivity. *Nature,* 1983; 306:686-688.

Woolf, C.J. (1992). Excitability changes in central neurons following peripheral damage: role of central sensitization in the pathogenesis of pain. In Willis, W.D. (Ed.), *Hyperalgesia and Allodynia* (pp. 221-243). New York: Raven Press.

Woolf, C.J. Windup and central sensitization are not equivalent. Editorial. *Pain,* 1996; 66:105-108.

Woolf, C.J. and King, A.E. Subthreshold components of the cutaneous mechanoreceptive fields of dorsal horn neurons in the rat lumbar spinal cord. *Journal of Neurophysiology,* 1989; 62:907-916.

Woolf, C.J., Shortland, P. and Coggeshall, R.E. Peripheral nerve injury triggers central sprouting of myhelinated afferents. *Nature,* 1992; 355:75-78.

Wu, J., Fang, L., Lin, Q. and Willis, W.D. C-fos expression is induced by increased nitric oxide release in rat spinal cord dorsal horn. *Neuroscience,* 2000; 96:351-357.

Wu, J., Lin, Q. and Willis, W.D. Increased phosphorylation of cAMP-responsive element-binding protein (CREB) following nitric oxide release in rat spinal cord. *Neuroscience Abstracts*, 1999.

Wu, J., Lin, Q., McAdoo, D.J. and Willis, W.D. Nitric oxide contributes to central sensitization following intradermal injection of capsaicin. *Neuroreport,* 1998; 9:589-592.

Yaksh, T.L. Behavioral and autonomic correlates of tactile evoked allodynia produced by spinal glycine inhibition: effects of modulatory receptor systems and excitatory amino acid antagonists. *Pain,* 1989; 37:111-123.

Yoon, Y.W., Na, H.S. and Chung, J.M. Contributions of injured and intact afferents to neuropathic pain in an experimental rat model. *Pain,* 1996; 64:27-36.

7 NOXIOUS STIMULUS-INDUCED PLASTICITY IN SPINAL CORD DORSAL HORN: EVIDENCE AND INSIGHTS ON MECHANISMS OBTAINED USING THE FORMALIN TEST

Terence J. Coderre
Director, Pain Mechanisms Laboratory
Clinical Research Institute of Montreal
Quebec, Canada, H2W 1R7

7.1 INTRODUCTION

After tissue or nerve injury, animals develop an increased sensitivity to noxious and non-noxious cutaneous stimuli, paralleling the development of hyperalgesia and allodynia in humans. While these changes depend partly on sensitization in peripheral tissues, there is also evidence that the increased sensitivity depends on sensitization within the central nervous system (CNS). Central sensitization after tissue or neural injury has been defined as a change in excitability of CNS neurons which is triggered by and outlasts nociceptive afferent inputs (Woolf and Chong, 1993). There is evidence indicating that persistent nociceptive responses to chemical stimulation with formalin depend on a sensitization of spinal cord neurons, which occurs in the early period after injury. Indeed, the formalin test has been used now for 10 years as an animal model of noxious stimulus-induced plasticity. In this paper, I will debate the evidence supporting the view that the formalin test reflects a sensitization of spinal cord neurons. Additionally, I will discuss some of the potential mediators that lead to the development of formalin injury-induced plasticity in spinal cord.

7.2 THE FORMALIN TEST AS A MODEL OF SPINAL NEURO-PLASTICITY

Pre-emptive Versus Post-Treatment

The possibility that events occurring early after a peripheral injury can lead to long-term changes in nociceptive behaviors has revived the notion, first proposed by Crile (1913), that it may be possible to pre-empt persistent pain by protecting the nervous system from the central sensitization associated with injury. The idea that persistent pain behaviors can be reduced by blocked the neural barrage associated with a peripheral injury has been referred to as pre-emptive analgesia. The formalin test has been used as a model of injury-induced central sensitization (Coderre, et al., 1990), and as an animal model for studying the potential usefulness of pre-emptive analgesia (Abram and Yaksh, 1993; Abram and Yaksh, 1994; Goto, et al., 1994a; Goto, et al., 1994b). Subcutaneous injection of dilute formalin into a rat's paw produces a biphasic response including an early intense response in the first 5 min, and a later moderate response that is expressed from 20 to at least 60 min after injection (Dubuisson and Dennis, 1977). The nociceptive response to formalin is matched by a corresponding biphasic increase in the activity of dorsal horn neurons after such injection (Dickenson and Sullivan, 1987).

It has been demonstrated that i.t. administration of either lidocaine (Coderre, et al., 1990; Abram and Yaksh, 1994) or opioids (Abram and Yaksh, 1993; Dickenson and Sullivan, 1987) substantially reduces behavioral and dorsal horn neuron responses to subcutaneous formalin, if they are administered prior to, but not immediately after, the early phase of the formalin response. This suggests that neural activity generated during the early phase of the formalin response is capable of producing changes in CNS function, which in turn influence nociceptive processing during the late phase. The ability of the pre-injury treatment with i.t. lidocaine or opioids to suppress the late phase response to formalin has been described as an animal model of pre-emptive analgesia, since the pretreatment apparently pre-empts the central sensitization which contributes to persistent nociceptive behaviors.

A specific role for spinal cord sensitization after hindpaw formalin injection is also indicated since lidocaine-induced block of the lumbar, but not the thoracic spinal cord, reduces the persistent nociceptive responses (Coderre, et al., 1994). Thus, the effects of lidocaine depend on the suppressing of activity of neurons intrinsic to the spinal level at which inputs arrive from the hindpaw, and not simply by the brief interruption of transmission to supraspinal structures. A lumbar locus for spinal sensitization contributing to the persistent responses to formalin is also supported by the data obtained from spinalized rats. Thus, we have found that persistent nociceptive reflexive behaviors (hindpaw flinching) can be demonstrated in rats chronically spinalized at the T6 level, indicating that persistent responses to formalin injection are abolished in spinalized rats. However, these effects were obtained for rats which were acutely spinalized (i.e., 48 hrs), as opposed our study where rats were chronically spinalized (i.e., > 21 days). It is possible that the trauma associated with acute spinal transection could have depressant influences on spinal somatosensory transmission.

The formalin test has been extensively used as an animal model for testing the potential of agents for pre-emptive analgesia. Several investigators have assessed whether it is possible to pre-empt persistent nociceptive responses in the formalin test using pretreatment with local or general anesthetics. Persistent nociceptive behaviors are significantly reduced by intravenous pretreatment with local anesthetics (Abram and Yaksh, 1994), or the general anesthetics propofol (Goto, et al., 1994b), pentobarbital (O'Connor and Abram, 1995) or alphaxalone (Gilron and Coderre, 1996) and by the inhalation of the general anesthetics halothane, enflurane, isoflurane, desflurane, or nitrous oxide (O'Connor and Abram, 1995).

Several investigations indicate that various treatments including opiates (Abram and Yaksh, 1993; Dickenson and Sullivan, 1987), or N-methyl-D-aspartate receptor (Haley, et al., 1990; Coderre and Melzack, 1992a and b; Yamamoto and Yaksh, 1992; Millan and Sequin, 1993; Eisenberg, et al., 1993) and neurokinin-1 receptor (Yamamoto and Yaksh, 1991; Birch, et al., 1992; Berge, et al., 1993; Yashpal, et al., 1993) antagonists, are often more effective at reducing nociceptive responses to formalin when given as pretreatments, as opposed to post-treatments. For example, the NMDA antagonist MK-801 significantly reduces nociceptive behaviors following formalin injury of a rat's hindpaw if administered prior to, but not following, the early phase response to formalin (Coderre and Melzack, 1992; Yamamoto and Yaksh, 1992b). Furthermore, the effects of MK-801 pretreatment are much more pronounced on the late phase responses to formalin, than on early phase responses, either on behavioural measures of nociception (Coderre and Melzack, 1992 a and b; Yamamoto and Yaksh, 1992), or on electrophysiological measures of dorsal horn neuronal activity (Haley, et al., 1990b).

Influence of Formalin Concentration

Although there is evidence that formalin-induced nociception is dramatically affected by pre-emptive analgesic treatments, this evidence is not unanimous (Haley, et al., 1990; Dallel, et al., 1995; Abbadie, et al., 1997; Taylor, 1997). We have shown, however, that part of this discrepancy can be explained by difference in the concentration of formalin that is used. For example, we (Yasphal, et al., 1996) have shown that the ability of i.t. lidocaine to pre-empt post-injury nociception in the formalin test was lost as the concentration of formalin was increased from 2.5 to either 3.75 or 5.0%. A strong pre-emptive effect (i.e., a significant reduction in nociceptive scores) of spinal lidocaine was obtained in rats given 2.5% formalin. This pre - emptive effect was lost (resulting in significantly higher nociceptive scores) in rats given 3.75 and 5.0% formalin. Furthermore, it was shown that injection of 3.75 and 5.0% formalin produced a significant degree of inflammation (plasma extravasation). In contrast, the degree of inflammation produced by 2.5% formalin was not significantly different from that produced by the same volume of saline, and was only slightly, but not significantly higher than no injection at all (Yasphal, et al., 1996). Thus, the pre-emptive effects of i.t. lidocaine are greatest when there is little or no inflammation, and decrease directly with increases in peripheral inflammation.

The suggestion that responses to high concentrations of formalin depend more on peripheral inflammation than central sensitization may explain the recent finding that there is no difference between pre- and post-treatment with i.t. lidocaine or excitatory amino acid (Chapman, et al., 1994) antagonists on the dorsal horn neuronal responses to a peripheral injection of 5.0% formalin to rats' toes, and why post-treatment with the NMDA antagonist AP5 produced a significant reduction in nociceptive responses to hindpaw injection of 10% formalin in mice (Murray, et al., 1991). In addition, nociception and late phase dorsal horn neuronal responses to 5.0% formalin are significantly reduced by local anesthesia of the injected area at the time of testing (Dickenson and Sullivan, 1987; Abbadie, et al., 1997), but not by a prior local anesthesia of the injected area during the early phase (Haley, et al., 1990). Furthermore, the magnitude of flinching and cardiovascular responses in the late phase of the response to 5.0% formalin is unaffected by i.v. treatment with remifentanil during the early phase (Abbadie, et al., 1997; Taylor, et al., 1997). This is consistent with the hypothesis that the persistent (late phase) nociception produced by a higher concentration of formalin is much more dependent on the ongoing afferent barrage associated with significant peripheral inflammation during the late phase. The mechanisms responsible for generating nociceptive behavior after administration of a low concentration of formalin differ markedly from the mechanisms present following a high concentration of formalin. Thus, it is important to pay careful attention to the concentration of formalin that the investigators have used, when comparing the results of different studies using the formalin test.

Influence of Anesthetics on Formalin-Induced Fos Protein Expression

Sensory inputs (Hunt, et al., 1987), in particular nociceptive inputs (Bullitt, et al., 1989), have been reported to produce increased expression of c-fos in the mammalian spinal dorsal horn. It has been suggested there may be a parallel between the ability of anesthetic and analgesic agents to pre-emptively inhibit nociceptive behaviors and to suppress Fos expression in spinal cord dorsal horn (Munglani, et al., 1993). This conclusion was based on the findings of Sun, et al. (1996) who examined the effects of inhalation anesthetics and fentanyl on both nociceptive responses and Fos expression in spinal cord following hindpaw injection of formalin. They observed that when inhalation anesthetics (2% halothane, 75% nitrous oxide or both) are used to block nociceptive responses in the first phase of the formalin test, they produce very little effect on either nociceptive responses in the second phase of the formalin test or Fos expression in spinal cord dorsal horn. When fentanyl (100 μg/kg, i.p.) is used to reduce first phase responses, however, both second phase nociceptive responses and formalin-induced spinal cord Fos expression is suppressed. Thus, both the persistent nociceptive responses and spinal cord Fos expression following formalin injection can be pre-empted by pretreatment with an opioid analgesic, but not with general anesthetics. We (Yashpal, et al., 1998) have provided evidence that spinal anesthetics can also produce consistent pre-emptive effects by demonstrating that pretreatment with i.t. lidocaine suppresses both second phase nociceptive responses and Fos expression induced by formalin.

Although we have shown that i.t. lidocaine pretreatment reduced both formalin-induced nociception and Fos expression; we also found that post-treatment with i.t. lidocaine reduced Fos expression, but not nociceptive responses to formalin (Yashpal, et al., 1998). These results parallel those of Tokunaga, et al. (1995) who found that post-treatment with 3% prilocaine (s.c. to the ankle) produced a similar suppression of formalin-induced Fos expression in dorsal horn as did pretreatment with 4% lidocaine. Thus, local anesthetics have differing effects on behavior and Fos expression depending on their use as pre- or post-treatments. Our recent findings suggest that may also be the case for various i.v. anesthetics and analgesics. Thus, we have recently shown that while i.v. ketamine and alphaxalone produce pre-emptive analgesia in the formalin test, the same treatments do not pre-emptively reduce formalin-induced Fos expression in spinal cord (Gilron and Coderre, 1999a). Alternatively, propofol and alfentanil produce a pre-emptive block of Fos expression which is not matched by pre-emptive analgesia in the formalin test (Gilron and Coderre, 1999b). These findings reveal an uncoupling of the behavioral and Fos protein responses to formalin. It should be noted that others (Hamalainen, et al., 1996) have also reported a dissociation of analgesics on pain behaviors and Fos expression in the spinal cord dorsal horn.

In our study with lidocaine, the finding that both persistent nociceptive behaviors and Fos expression were suppressed by i.t. lidocaine pretreatment, supports the hypothesis that nociception in the second phase of the formalin test is dependent on increases in central hyperexcitability generated during the first phase. However, since both Fos expression and second phase nociceptive responses to formalin were not completely eliminated by i.t. lidocaine pretreatment, this suggests that peripheral inputs also contribute to the expression of nociception in the second phase of the formalin test. Since i.t. lidocaine pretreatment would substantially reduce spinal activation during the first phase, residual Fos expression and second phase nociceptive behaviors are mostly likely explained by peripheral inputs during the second phase. A contribution of ongoing peripheral inputs to second phase formalin responses is consistent with previous results indicating that behavioral (Coderre, et al., 1990; Dallel, et al., 1995; Taylor, et al., 1995), spinal cord dorsal horn neuronal (Dickenson and Sullivan, 1987), Fos protein expression (Abbadie, et al., 1997) and reflex cardiovascular (Taylor, et al., 1995) responses to formalin are significantly reduced or eliminated by infiltration of the formalin-injected site with local anesthetics at the time of testing during the second phase. The importance of ongoing peripheral inputs is also suggested since formalin injections have been shown to produce biphasic activity in both C- and A-delta- primary afferent fibers (McCall, et al., 1996; Puig and Sorkin, 1996), although others (Abram, et al., 1996) have suggested that primary afferent fiber activity in the second phase is considerably less than in the first phase.

7.3 MEDIATORS OF SPINAL NEUROPLASTICITY AFTER FORMALIN INJURY: NEUROTRANSMITTERS

Although ongoing peripheral inputs likely contribute to formalin-induced nociception, the above data indicate that spinal neuroplasticity also plays a role. Formalin injury initiates its central neural consequences by triggering the release of various neurotransmitters into the dorsal horn of the spinal cord. Among the most critical of these neurotransmitters are the neuropeptide, including substance P and the excitatory amino acids, including glutamate and aspartate.

Neuropeptides

Formalin injury has been found to increase substance P-immunoreactivity in perfusates from the spinal cord intrathecal space (Kuraishi, et al., 1989; McCarson and Goldstein, 1991) as well as increasing substance P- (Kantner, et al., 1985; Holland and Goldstein, 1990; Zhang, et al., 1994), CGRP-, somatostatin- and neurotensin-immunoreactive materials in the superficial layers of the dorsal spinal cord (Zhang, et al., 1994). Formalin injury also increases the level of preprotachykinin (Noguchi, et al., 1988; McCarson and Krause, 1994) and neurokinin B mRNA (McCarson and Krause, 1995) in the dorsal root ganglion, and NK-1 and NK-3 receptor mRNA in the spinal dorsal horn (McCarson and Krause, 1994). Recently, it has been shown that there is an increase in the internalization of substance P receptors at both 8 and 60 min after formalin injury (Honor, et al., 1999). In support of this, we have also shown that there is a biphasic decrease in the binding of exogenous substance P to the superficial dorsal horn of spinal cord sections taken from rats with a formalin injury (Henry, et al., 1999). These results suggest that there is a biphasic release of substance P after formalin injury, which corresponds to the biphasic nociceptive responses seen in formalin-injured animals.

Further evidence indicates that formalin-induced nociception is attenuated by the substance P or NK-1 antagonists. Tonic nociceptive responses to formalin are reduced by systemic pretreatment with nonpeptide NK-1 antagonists (Garret, et al., 1991; Yamamoto and Yaksh, 1991; Birch, et al., 1992; Berge and Ståhlberg, 1993; Yamamoto, et al., 1993; Yashpal, et al., 1993; Seguin, et al., 1995). Pretreatment with a nonpeptide NK-1 antagonist also reduces formalin-induced Fos expression in spinal cord dorsal horn (Chapman, et al., 1996). Formalin-induced nociception or increases in dorsal horn neuronal activity are also reduced by intrathecal injection of both peptide (Ohkubo, et al., 1990; Murray, et al., 1991; Sakurada, et al., 1993a; 1995) and nonpeptide (Chapman and Dickenson, 1993; Sakurada, et al., 1993b; Traub, 1996; Henry, et al., 1999) substance P antagonists. While some investigators have found that pretreatment with substance P antagonists affect only late phases responses to formalin (Yamamoto and Yaksh, 1991; Birch, et al., 1992; Berge and Ståhlberg, 1993; Yamamoto, et al., 1993; Yashpal, et al., 1993; Seguin, et al.,1995; Traub, 1996) others have suggested it inhibits only the early phase (Ohkubo, et al., 1990), or that it inhibits both nociceptive phases (Sakurada, et al., 1993a,b; 1995; Chapman and Dickenson, 1993). Intrathecal post-treatment with either peptide (Murray, et al., 1990) or nonpeptide (Henry, et al., 1999) substance P antagonist have also been found to effectively reduce formalin-induced nociception, although this effect is not supported by Yamamoto and Yaksh (1991) or Yamamoto, et al. (1993). Overall the data support the suggestion that substance P contributes to formalin-induced nociception over a prolonged period, but that the best inhibition is obtained in the late phase after pretreatment with substance P antagonists.

Excitatory Amino Acids

Formalin injury also increases the release of glutamate and aspartate from the dorsal horn of the spinal cord (Skilling, et al., 1988; Marsala, et al., 1995). Furthermore, evidence indicates that formalin-induced nociception is attenuated by the various excitatory amino acid antagonists. Persistent nociceptive responses induced by formalin injections in the hindpaw of mice and rats have been reduced with the spinal administration of competitive antagonists (Murray, et al., 1991; Coderre and Melzack, 1992; Näsström, et al., 1992; Millan and Sequin, 1994; Hunter and Singh, 1994; Kristensen, et al., 1994; Chaplan, et al., 1997). Formalin-induced nociception or dorsal horn neuronal activity is also attenuated by spinal or systemic administration of noncompetitive NMDA antagonists acting at the phencylclidine (PCP) (Elliot, et al., 1995; Näsström, et al., 1992; Millan and Sequin, 1993; 1994; Eisenberg, et al., 1993; Coderre and Melzack, 1992; Yamamoto and Yaksh, 1992; Vaccarino, et al., 1993; Chaplan, et al., 1997; Shimoyama, et al., 1997; Takano, et al., 2000) or glycine modulatory sites (Näsström, et al., 1992; Millan and Sequin, 1994; Hunter and Singh, 1994; Lutfy and Weber, 1996; Dickenson and Aydar, 1991; Chapman and Dickenson, 1995) In addition, formalin nociception is reduced by AMPA/kainate antagonists (Näsström, et al., 1992; Simmons, et al., 1998; Nishiyama, et al., 1999aandb), and to a lesser extent group I mGluR antagonists (Fisher and Coderre, 1996). A spinally administered NMDA antagonist has also been found to reduce the formalin injury-induced increase of Fos protein in spinal cord dorsal horn (Ji and Rupp, 1997; Tao and Zhao, 1998).

Early evidence suggested that NMDA antagonists significantly reduce nociceptive behaviors following formalin injury if administered prior to, but not following, the early phase response to formalin (Coderre and Melzack, 1992; Yamamoto and Yaksh, 1992; Vaccarino, et al., 1993), although see Murray, et al. (1990). Furthermore, the effects of pretreatment with NMDA antagonists are much more pronounced on the late phase responses to formalin, than on early phase responses, either on behavioural measures of nociception (Coderre and Melzack, 1992; Yamamoto and Yaksh, 1992; Millan and Seguin, 1993; Hunter and Singh, 1994; Shimoyama, et al., 1997; 1999; Takano, et al., 2000), or on electrophysiological measures of dorsal horn neuronal activity (Haley, et al., 1990), although also see Kristensen, et al. (1994). In contrast, AMPA/kainate antagonists reduce only early (Hunter and Singh, 1994) or both early and late phase formalin-induced nociceptive responses (Nishiyama, et al., 1999aandb). Once again as for substance P, the data support the suggestion that NMDA receptor activity contributes to formalin-induced nociception over a prolonged period, but that the best inhibition is obtained in the late phase after pretreatment with NMDA antagonists.

7.4 MEDIATORS OF SPINAL NEUROPLASTICITY AFTER FORMALIN INJURY: SECOND MESSENGERS AND PROTEIN KINASES

The above data suggest that both neuropeptides and EAAs contribute to spinal neuroplasticity affecting nociceptive responses to formalin. The manner by which these substances produce long-term effects has become evident over the last few years. It has been demonstrated that neuropeptides, such as substance P, and excitatory amino acids trigger alterations in membrane permeability and synaptic efficacy through interactions with second messengers and protein kinases, which phosphorylate membrane-bound proteins. Evidence indicates that increases in intracellular calcium (Ca2+), second messengers and kinases including, phospholipases, nitric oxide and protein kinase C play a significant role in the development of spinal plasticity following formalin injury.

Intracellular Calcium

Evidence suggests that noxious stimulation may produce an increase of intracellular Ca2+ in nociceptive neurons, which influences the excitability of the cell. Neurotransmitters released in response to noxious stimulation are known to affect the intracellular levels of Ca2+. Glutamate and aspartate stimulate the influx of Ca2+ through NMDA receptor-operated channels (MacDermott, et al., 1986). Substance P produces an elevation in intracellular Ca2+ by mobilizing its release from intracellular stores (Womack, et al., 1988), while both substance P (Womack, et al., 1989) and CGRP (Oku, et al., 1988) increase Ca2+ influx through voltage-gated Ca2+ channels.

Over the last few years it has been shown that nociceptive response to formalin are affected by manipulating the levels of intracellular calcium in the spinal dorsal horn. We have previously shown that nociceptive responses to formalin are reduced by intrathecal administration of L-type Ca2+ channel antagonists and a Ca2+ chelator which decrease intracellular Ca2+, and are increased by spinal administration of a Ca2+ ionophore or Ca2+ agonist which increase intracellular Ca2+ (Coderre and Melzack, 1992b). Subsequently, it has been shown that formalin-induced nociception is reduced by trivalent cations which usurp extracellular Ca2+ and conopeptides or ziconotide which block N- and P-type Ca2+ channels (Malmberg and Yaksh, 1994b; 1995b; Bowersox, et al., 1996; Nakanishi, et al., 1999; Wang, et al. 2000). Furthermore, formalin-induced nociception is reduced by inhibitors of calmodulin (Menendez, et al., 1996), a Ca2+ - binding protein which serves to activates several intracellular enzymes.

Nitric Oxide

When Ca2+ binds to calmodulin this leads to the activation of nitric oxide (NO) synthase. NO synthase converts L-arginine to NO, and the released NO stimulates the production of cGMP from soluable guanylate cyclase (Synder, 1992). NO synthase inhibitors and inhibitors of soluable guanylate cyclase produce analgesic effects in various nociceptive tests, including the formalin test. Formalin injury produces an increase in the number of spinal neurons expressing NO synthase (Lam, et al., 1996), and increases in spinal citrulline (a byproduct of NO synthase activity) determined by microdialysis (Malmberg and Yaksh, 1995c). Formalin injury also produces an increase in the cGMP-dependent protein kinase II protein expression in spinal cord, which is blocked by NOS inhibitors (Tao, et al., 2000). It has shown that formalin injury produces an increase in spinal NOS activity as detected by increased NOS-immunoreactivity and NADPH-diaphorase reaction in spinal cord dorsal horn (Herdegen, et al., 1994). Importantly, neurons expressing a NADPH-d reaction were determined to be in close spatial relationship to fibers expressing substance P and CGRP, and were co-localized with c-fos mRNA, as well as mRNA for various other proto-oncogenes. This suggests there is a functional relationship between the expression of proto-oncogenes, such as c-fos, and nitric oxide in spinal cord neurons activated by formalin injury. The close relationship between substance P neurons and those expressing c-fos was confirmed by Tao and Zhao (1997), is also supported by a subsequent study demonstrating a high degree of neurons in lamina I of the spinal cord which co-localize NK-1 receptors and Fos protein immunoreactivity after formalin injury (Doyle and Hunt, 1999). Neurons in the spinal cord dorsal horn that are Fos positive after formalin injury have also been found to co-localize with NMDA1 receptors (almost all) and Glu2,3 receptors (94%), as well as NOS (14%) (Leong, et al., 2000).

Moore, et al. (1991) was the first to demonstrate that systemic treatment with L-NG-nitro arginine methyl ester (L-NAME) produced analgesia in the formalin test. Of interest, L-NAME pretreatment produced more significant analgesia effects in the late phase of the formalin test than in the early phase, suggesting that NO may be involved in the development of persistent pain associated with central sensitization. In addition, Haley, et al. (1992) demonstrated that intrathecal L-NAME reduced the persistent increase in dorsal horn neuron activity following formalin injury. Subsequently, several investigators have shown that spinal administration of NOS inhibitors produce a significant reduction of formalin-induced nociception (Malmberg and Yaksh, 1993; Yamamoto, et al., 1993; Seguin, et al., 1995; Sakurada, et al., 1996; Roche, et al., 1996), with a majority of them indicating greater effects during the late phase. Systemic (Chapman, et al., 1995) or intrathecal (Roche, et al., 1996) NOS inhibitors, as well as an inhibitor of cGMP protein kinase II (Tao, et al., 2000), produce a reduction in formalin-induced Fos protein expression. In our own studies (Coderre and Yashpal, 1994), we have demonstrated that intrathecal treatment with sodium nitroprusside, an NO donor which increases NO levels, produces an enhancement of nociceptive responses to formalin, while L-NAME produces antinociceptive effects, as well as a reduction in the hyperalgesic effects of glutamate, in the formalin test. Others have recently supported the finding that formalin-induced nociception is enhanced after intrathecal treatment with an NO donor (Shibuta, et al., 1996).

Arachidonic Acid and Its Metabolites

In addition to NO, there is an increased production of spinal arachidonic acid in response to noxious stimulation. Noxious stimulus-induced increases in intracellular Ca2+ produce a phospholipase A2-catalyzed stimulation of arachidonic acid (Dumuis, et al., 1990). Further increases in arachidonic acid occur following the breakdown of diacyclygerol (Gammon, et al., 1989), which is produced following stimulation of phospholipase C with activity at G-protein coupled metabotropic receptors. A role of intracellular arachidonic acid in the development of neuronal plasticity is consistent with the finding that arachidonic acid regulates a number of ionic conductances either directly (Kim and Clapham, 1989), or following its metabolism into various eicosanoids (Kuriashi, et al., 1989).

Evidence suggests that arachidonic acid metabolites in spinal cord contribute to nociceptive processing after formalin injury. First, formalin injury produces an increased release of spinal prostaglandin (PG) E2, a cylo-oxygenase product of arachidonic acid metabolism (Malmberg and Yaksh, 1995aandc; Hua, et al., 1997; Muth-Selbach, et al., 1999). Evidence also indicates that spinal intrathecal administration of non-steroidal anti-inflammatory drugs (NSAIDs) produces analgesia in the formalin test (Malmberg and Yaksh, 1992; 1994aandb; Malmberg, et al., 1994; Yamamoto, et al., 1996; Dirig, et al, 1997). Like NO synthase inhibitors, the intrathecal NSAIDs are more effective at inhibiting the late phase of the formalin response than they are at inhibiting the early phase. Once again this suggests that prostanoids may be involved in the development of persistent pain associated with central sensitization. In our own studies (Coderre and Yashpal, 1994), we have demonstrated that inhibition of arachidonic acid with dexamethasone produces antinociceptive effects, as well as a reduction in the hyperalgesic effects of glutamate, in the formalin test. In contrast, the spinal administration of either arachidonic acid (Coderre and Yashpal, 1994) or PGE1 (Takano, et al., 1999) produce an enhancement of formalin-induced nociception.

Protein Kinase C

Evidence also implies that noxious stimulus-induced plasticity in the formalin test may depend, in part, on the phospholipase C (PLC) second messenger system. Activity at either NK-1 receptors by SP (Mantyh, et al., 1984), or at metabotropic EAA receptors by glutamate and aspartate (Sugiyama, et al., 1987), stimulates the hydrolysis of inositol phospholipids by activating a polyphosphoinositide-specific PLC. PLC is an enzyme, which catalyzes the hydrolysis of polyphosphatidylinositol into the intracellular messengers inositol trisphosphate (IP3) and diacylglycerol (DAG). It has been shown that neomycin (an agent which inhibits PLC activity) produces a substantial reduction in nociceptive responses during the formalin test (Coderre, 1992). Since inhibition of PLC with neomycin produces a significant reduction in formalin nociceptive responses, it is assumed that IP3 and DAG play a significant role in formalin-induced plasticity. Following its production, IP3 stimulates the release of Ca2+ from internal stores; on the other hand DAG stimulates the translocation and activation of protein kinase C (PKC). When activated by DAG, PKC phosphorylates specific substrate proteins that contribute to various cellular processes, including neurotransmitter release and transduction (Nishizuka, 1986).

Recently, Gerber, et al. (1989) have shown that activators of PKC enhance the basal and evoked release of glutamate and aspartate in the spinal cord slice, as well as the depolarizing responses of dorsal horn neurons to exogenous glutamate and NMDA. Furthermore, Chen and Huang (1992) have demonstrated that PKC increases NMDA-activated currents in isolated trigeminal cells by increasing the probability of channel openings and by reducing the voltage-dependent Mg2+ block of NMDA-receptor channels. The resultant increased activity at the NMDA receptor would permit more Ca2+ ions to enter the cell, raising intracellular Ca2+, and so further increasing PKC activity, and its effects on NMDA receptor channels. This positive feedback loop may be important for the induction or maintenance of sensitization in central neurons.

We have previously shown that formalin injury produces an increase in membrane-bound protein kinase C in dorsal horn as determined by the binding of 3H-phorbol 12,13 dibutyrate (Yashpal, et al., 1995). Behavioural studies indicate that nociceptive responses to formalin injury are suppressed following intrathecal treatment with inhibitors of PKC (Coderre, 1992; Yashpal, et al., 1995; Nakanishi, et al., 1999), and enhanced after treatment with phorbol esters or SC-10 which stimulate PKC (Coderre, 1992, Coderre and Yashpal, 1994). These results are consistent with a contribution of PKC to the persistent nociceptive response elicited by subcutaneous injection of formalin. Furthermore, the enhancement of formalin-induced nociception after intrathecal administration with substance P or glutamate is reversed by a PKC inhibitor (Coderre and Yashpal, 1994)

7.5 CONCLUSIONS

In addition to peripheral changes, tissue injury produced by formalin injection leads to changes in spinal neuronal function, which contribute to persistent nociception. It is proposed that central sensitization which occurs early after formalin injection contributes to the persistent nociceptive responses to formalin. Since pretreatments often only affect the late phase responses, and not the early phase, it is expected that these treatments do not affect acute nociception, but inhibit its transition into hypersensitivity, which contributes to persistent nociception. Various treatments are able to pre-empt the development of hyperexcitability and reduce persistent nociceptive behaviors in the formalin test, although this effect is dependent on the concentration of the formalin injected. When the formalin concentration is high, it is expected that ongoing peripheral inputs from the inflamed tissue obscure the ability to detect a contribution of spinal plasticity to the late phase of the formalin test.

The evidence presented suggests that neuropeptides, excitatory amino acids and intracellular messengers to which they are coupled play a significant role in the development of noxious stimulus-induced plasticity and the persistent nociception it causes in the formalin test. It is proposed that intense noxious stimulation associated with formalin injury leads to a release of neuropeptides and EAA transmitters within spinal dorsal horn. Following the activation their receptors, these transmitters produce an increase in intracellular Ca2+, NO, phospholipase A2 and phospholipase C, leading to the activation or production of soluable guanylate cyclase, arachidonic acid and polyphosphoinosites (IP3, DAG) (Garthwaite, et al., 1988; Dumuis, et al., 1988; Sladeczek, et al., 1985) which in turn trigger the release or production of cGMP, eicosanoids and protein kinase C, respectively (Southam, et al., 1991; Manzoni, et al., 1990; Gammon, et al., 1989). The activation of each of these second messengers is proposed to produce intracellular changes including the phosphorylation of substrate proteins and membrane-bound receptor channels (Mayer, et al., 1991). Importantly, it has been suggested that Ca2+ influx, as occurs following NMDA receptor activation, is required to transform protein kinase C to its activated state (Rasmussen, 1986). The activated protein kinase C would not only contribute to membrane phosphorylation, but along with Ca2+, would trigger the production of immediate-early genes such as c-fos (Naranjo, et al., 1991). This chain of events would lead to prolonged alterations in the neurons' responses to further stimulation, and would thus account for the ability of the intense noxious stimulation associated with formalin injury to produce persistent nociceptive responses.

7.6 REFERENCES

Abbadie, C., Taylor, B.K., Peterson, M.A., Basbaum, A.I. Differential contribution of the two phases of the formalin test to the pattern of c-fos expression in the rat spinal cord: studies with remifentanil and lidocaine. *Pain,* 1997; 69:101-10.

Abram, S.E., Dean, C., O'Connor, T.C. Peroneal afferent nerve discharges underlying the behavioral response to the formalin test. *Regional Anaesthesie,* 1996; 21:226-33.

Abram, S.E. Yaksh, T.L. Morphine, but not inhalation anesthesia, blocks post-injury facilitation: The role of preemptive suppression of afferent transmission. *Anesthesiology,* 1993; 78:713-21.

Abram, S.E., Yaksh, T.L. Systemic lidocaine blocks nerve injury-induced hyperalgesia and nociceptor-driven spinal sensitization in the rat. *Anesthesiology,* 1994; 80:383-91.

Berge, O-G., Ståhlberg, M. Is the selective non-peptide NK-1 receptor antagonist CP96, 345 a peripherally acting analgesic. *Regulatory Peptides,* 1993; 46:430-2.

Birch, P.J., Harrison, S.M., Hayes, A.G., Rogers, H., Tyers, M.B. The non-peptide NK1 receptor antagonist, (±)-CP-96,345, produces antinociceptive and anti-oedema effects in the rat*British Journal of Pharmacology,* 1992; 105:508-10.

Bowersox, S.S., Gadbois, T., Singh, T., Pettus, M., Wang, Y.X., Luther, R.R. Selective N-type neronal voltage-sensitive calcium channel blocker, SNX-111, produces spinal antinociception in rat models of acute, persistent and neuropathic pain. *Journal of Pharmacology and Experimental Therapeutics,* 1996; 279:1243-49.

Bullitt, E. Induction of c-fos-like protein within the lumbar spinal cord and thalamus of the rat following peripheral stimulation. *Brain Research,* 1989; 493:391-397.

Chaplan, S.R., Malmberg, A.B., Yaksh, T.L. Efficacy of spinal NMDA receptor antagonism in formalin hyperalgesia and nerve injury evoked allodynia in the rat. *Journal of Pharmacology and Experimental Therapeutics,* 1997; 280: 829-38.

Chapman, V., Buritova, J., Honore, P., Besson, J.M. 7-Nitro-indazole, a selective inhibitor of neuronal nitric oxide synthase, reduces formalin evoked c-Fos expression in dorsal horn neurons of the rat spinal cord. *Brain Research,* 1995; 697:258-61.

Chapman, V., Buritova, J., Honore, P., Besson, J.M.. Physiological contributions of neurokinin 1 receptor activation, and interactions with NMDA receptors, to inflammatory-evoked spinal c-Fos expression. *Journal of Neurophysiology,* 1996; 76:1817-27.

Chapman, V., Dickenson, A.H. The effect of intrathecal administration of RP67580, a potentneurokinin 1 antagonist on nociceptive transmission in the rat spinal cord. *Neuroscience Letters,* 1993; 157:149-52.

Chapman, V., Dickenson, A.H. Time-related roles of excitatory amino acid receptors during persistent noxiously evoked responses of rat dorsal horn neurones. *Brain Research,* 1995; 703:45-50.

Chapman, V., Haley, J.E., Dickenson, A.H. Electrophysiologic analysis of preemptive effects of spinal opioids on N-methyl-D-aspartate receptor-mediated events. *Anesthesiology,* 1994; 81:1429-35.

Chen, L., Huang, L-YM. Protein kinase C reduces Mg2+ block of NMDA-receptor channels as a mechanism of modulation. *Nature,* 1992; 356:521-23.

Coderre, T.J. Contribution of protein kinase C to persistent nociception following tissue injury in rats. *Neuroscience Letters,* 192; 140:181-84.

Coderre, T.J., Melzack, R. The contribution of excitatory amino acids to central sensitization and persistent nociception after formalin-induced tissue injury. *Journal of Neuroscience,* 1992a; 12:3665-70.

Coderre, T.J., Melzack, R. The role of NMDA receptor-operated calcium channels in persistent nociception after formalin-induced tissue injury. *Journal of Neuroscience,* 1992b; 12:3671-3675.

Coderre, T.J., Vaccarino, A.L., Melzack, R. Central nervous system plasticity in the tonic pain response to subcutaneous formalin injection. *Brain Research,* 1990; 535:155-58.

Coderre, T.J., Yashpal, K. Intracellular messengers contributing to persistent nociception and hyperalgesia induced by L-glutamate and substance P in the rat formalin pain model. *European Journal of Neuroscience,* 1994; 6:1328-34.

Coderre, T.J., Yashpal, K., Henry, J.L. Specific contribution of lumbar spinal mechanisms to persistent nociceptive responses in the formalin test. *Neuroreport,* 1994; 5:1337-40.

Crile, G.W. *The kinetic theory of shock and its prevention through anoci-association (shockless operation). Lancet,* 1913; ii: 7-16.

Dallel, R., Raboisson, P., Clavelou, P., Saade, M., Woda, A. Evidence for a peripheral origin of the tonic nociceptive response to subcutaneous formalin. *Pain,* 1995; 61:11-16.

Dickenson, A.H., Aydar, E.. Antagonism at the glycine site on the NMDA receptor reduces spinal nociception in the rat. *Neuroscience Letters,* 1991; 121:263-66.

Dickenson, A.H., Sullivan, A.F. Subcutaneous formalin-induced activity of dorsal horn neurons in the rat: differential response to an intrathecal opiate administration pre or post formalin. *Pain,* 1987; 30:349-60.

Dirig, D.M., Konin, G.P., Isakson, P.C., Yaksh, T.L. Effect of spinal cyclooxygenase inhibitors in rat using the formalin test and in vitro prostaglandin E2. *European Journal of Pharmacology,* 1997; 331:155-60.

Doyle, C.A., Hunt, S.P. Substance P receptor (neurokinin-1)-expressing neurons in lamina 1 of the spinal cord encode for the intensity of noxious stimulation: a c-Fos study in rat*Neuroscience,* 1999; 89:17-28.

Dubuisson, D., Dennis, S.G. The formalin test: a quantitative study of the analgesic effects of morphine, meperidine, and brain stem stimulation in rats and cats. *Pain,* 1977; 4:161-74.

Dumuis, A., Pin, J-P., Oomagari, K., Sebben, M., Bockaert, J. Arachidonic acid released from striatal neurons by joint stimulation of ionotropic and metabotropic quisqualate receptors. *Nature,* 1990; 347:182-84.

Dumuis, A., Sebben, M., Haynes, L., Pin, J-P., Bockaert, J. NMDA receptors activate the arachidonic acid cascade system in striatal neurons. *Nature,* 1988; 336:68-70.

Eisenberg, E., Vos, B.P., Strassman, A.M. The NMDA antagonist Memantine blocks pain behavior in a rat model of formalin-induced facial pain. *Pain,* 1993; 54:301-7.

Elliott, K.J., Brodsky, M., Hynansky, A.D., Foley, K.M., Inturrisi, C.E. Dextromethorphan suppresses both formalin-induced nociceptive behaviour and the formalin-induced increase in spinal cord c-fos mRNA. *Pain,* 1995;61: 401-9.

Fisher, K., Coderre, T.J. The contribution of metabotropic glutamate receptors (mGluRs) to formalin-induced nociception. *Pain,* 1996; 68:255-63.

Gammon, C.M., Allen, A.C.C., Morell, P. Bradykinin stimulates phosphoinositide hydrolysis and mobilization of arachidonic acid in dorsal root ganglion neurons.*Journal of Neurochemistry,* 1989; 53:95-101.

Garret, C., Carruette, A., Fardin, V., Moussaoui, S., Peyronel, J.F., Blanchard, J.C., Laduron, P.M. Pharmacological properties of a potent and selective nonpeptide substance P antagonist. *Proceedings of The National Academy of Sciences of The United States of America,* 1991; 88:10208-12.

Garthwaite, J., Charles, S.L., Chess-Williams, R. Endothelium-derived relaxing factors released on activation of NMDA receptors suggests role as intracellular messenger in the brain.*Nature,* 1988; 336:385-88.

Gerber, G., Kangrga, I., Ryu, P.D., Larew, J.S.A., Randic, M. Multiple effects ofphorbol esters in the rat spinal dorsal horn. *Journal of Neuroscience,* 1989; 9:3606-17.

Gilron, I., Coderre, T.J. Preemptive analgesic effects of steroid anesthesia with alphaxalone in the rat formalin test. Evidence for differential GABA(A) receptor modulation in persistent nociception. *Anesthesiology,* 1996; 84:572-9.

Gilron, I., Quirion, R., Coderre, T.J. Pre-versus postformalin effects of ketamine or large-dose alfentanil in the rat: discordance between pain behavior and spinal Fos-like immunoreactivity. *Anesthesia and Analgesia,* 1999a; 89:128-35.

Gilron, I., Quirion, R., Coderre, T.J. Pre- versus postinjury effects of intravenous GABAergic anesthetics on formalin-induced Fos immunoreactivity in the rat spinal cord. *Anesthesia Analgesia,* 1999b; 88:414-20.

Goto, T., Marota, J.J.A., Crosby, G. Nitrous oxide induces preemptive analgesia in the rat that is antagonized by halothane.*Anesthesiology,* 1994a; 80:409-16.

Goto, T., Marota, J.J.A., Crosby, G. Pentobarbitone, but not propofol, produces pre-emptive analgesia in the rat formalin model. *British Journal of Anaesthesia,* 1994b; 72:662-7.

Haley, J., Ketchum, S., Dickenson, A.H. Peripheral κ-opioid modulation of the formalin response: an electrophysiological study in the rat. *European Journal of Pharmacology,* 1990a; 191:437-46.

Haley, J.E., Sullivan, A.F., Dickenson, A.H. Evidence for spinal N-methyl-D-aspartate receptor involvement in prolonged chemical nociception in the rat. *Brain Research,* 1990b; 518:218-26.

Haley, J.E., Wilcox, G.L., Chapman PF. The role of nitric oxide in hippocampal long-term potentiation. *Neuron,* 1992; 8:211-16.

Hamalainen, M.M., Alila, A., Johansson, G., Pertovaara, A. Cocaine-induced effects on pain behavior and c-fos expression in the spinal dorsal horn of the rat. *Neuroscience Research Communication,* 1996; 19:67-74.

Herdegen, T., Rudiger, S., Mayer, B., Barvo, R. Expression of nitric oxide synthase and colocalisation with Jun, Fos and Krox transcription factors in spinal cord neurons following noxious stimulation of the rat hindpaw. *Brain Research. Molecular Brain Research,* 1994; 22:245-58.

Holland, L.N., Goldstein, B.D. Changes of substance P-like immunoreactivity in the dorsal horn are associated with the "phasic" behavioral response to a formalin stimulus. *Brain Research,* 1990; 537:287-92.

Honor, P., Menning, P.M., Rogers, S.D., Nichols, M.L., Basbaum, A.I., Besson, J.M., Mantyh, P.W. Spinal substance P receptor expression and internalization in acute, short-term, and long-term inflammatory pain states. *Journal of Neuroscience,* 1999; 19:7670-8.

Hua, X.Y., Calcutt, N.A., Malmberg, A.B. Neonatal capsaicin treatment abolishes formalin-induced spinal PGE2 release. *Neuroreport,* 1997; 8:2325-9.

Hunt, S.P., Pini, A., Evan, G. Induction of c-fos-like protein in spinal cordneurones following sensory stimulation. *Nature,* 1987; 328:632-4.

Hunter, J.C., Singh, L. Role of excitatory amino acid receptors in the mediation ofnociceptive response to formalin in the rat. *Neuroscience Letters,* 1994; 174: 217-221.

Ji, R.R., Rupp, F. Phosphorylation of transcription factor CREB in rat spinal cord afterformalin-induced hyperalgesia: relationship to c-fos induction. *Journal of Neuroscience,* 1997; 17:1776-85.

Kantner, R.M., Kirby, M.L., Goldstein, B.D. Increase in substance P in the dorsal horn during a chemogenic nociceptive stimulus. *Brain Research,* 1985; 338:196-9.

Kim, D., Clapham, D.E. Potassium channels in cardiac cells activated by arachidonic acid and phospholipids. *Science,* 1989; 224:1174-76.

Kristensen, J.D., Karlsten, R., Gordh, T., Berge, O-G. The NMDA antagonist 3-(2-carboxypiperazin-4-yl)propyl-1-phosphonic acid (CPP) hasantinociceptive effect after intrathecal injection in the rat. *Pain,* 1994; 56: 59-67.

Kuraishi, Y., Hirota, N., Sato, Y., Hanashima, N., Takagi, H., Satoh, M. Stimulus specificity of peripherally evoked substance P release from the rabbit dorsal horn in situ.*Neuroscience,* 1989; 30:241-50.

Lam, H.H., Hanley, D.F., Trapp, B.D., Saito, S., Raja, S., Dawson, T.M., Yamaguchi, H. Induction of spinal cord neuronal nitric oxidesynthase (NOS) after formalin injection in the rat hind paw. *Neuroscience Letters,* 1996; 210:201-4.

Leong, S., Liu, H., Yeo, J. Nitric oxide synthase and glutamate receptor immunoreactivity in the rat spinal trigeminal neurons expressing Fos protein after formalin injection. *Brain Research,* 2000; 855:107-15.

Lutfy, K., Weber, E.. Attenuation of nociceptive responses by ACEA-1021, a competitive NMDA receptor glycine site antagonist, in mice. *Brain Research,* 1996; 743:17-23.

MacDermott, A.B., Mayer, M.L., Westbrook, G.L., Smith, S.J., Barker, J.L. NMDA-receptor acyivation increases cytoplasmic calcium concentration in cultured spinal cord neurons.*Nature,* 1986; 321:519-22.

Malmberg, A.B., Rafferty, M.F., Yaksh, T.L. Antinociceptive effect of spinally delivered prostaglandin E receptor antagonists in the formalin test on the rat. *Neuroscience Letters,* 1994; 173:193-6.

Malmberg, A.B., Yaksh, T.L. Antinociceptive actions of spinal nonsteroidal anti-inflammatory agents on the formalin test in the rat. *Journal of Pharmacology and Experimental Therapeutics,* 1992; 263:136-46.

Malmberg, A.B., Yaksh, T.L. Spinal nitric oxide synthesis inhibition blocks NMDA-induced thermal hyperalgesia and produces antinociception in the formalin test in rats. *Pain,* 1993; 54:291-300.

Malmberg, A.B., Yaksh, T.L. Antinociception produced by spinal delivery of the S and Renantiomers of flurbiprofen in the formalin test. *European Journal of Pharmacology,* 1994a; 260:283.

Malmberg, A.B., Yaksh, T.L. Voltage-sensitive calcium channels in spinal nociceptive processing: blockade of N- and P-type channels inhibits formalin-induced nociception. *Journal of Neuroscience,* 1994b; 14:4882-90.

Malmberg, A.B., Yaksh, T.L. Cyclooxygenase inhibition and the spinal release of prostaglandin E2 and amino acids evoked by paw formalin injection: a microdialysis study in unanesthetized rats. *Journal of Neuroscience,* 1995a; 15:2768-76.

Malmberg, A.B., Yaksh, T.L. Effect of continuous intrathecal infusion of omega-conopeptides, N-type calcium-channel blockers, on behaviors and antinociception in the formalin and hot-plate test in rats. *Pain,* 1995b; 60:83-90.

Malmberg, A.B., Yaksh, T.L.. The effect of morphine on formalin-evoked behaviour and spinal release of excitatory amino acids and prostaglandin E2 usingmicrodialysis in conscious rats. *British Journal of Pharmacology,* 1995c; 114:1069-75.

Mantyh, P.W., Pinnock, R.D., Downes, C.P., Goedert, M., Hunt, S.P. Correlation between inositol phospholipid hydrolysis and substance P receptors in rat CNS. *Nature,* 1984; 309:795-97.

Manzoni, O.J.J., Finiels-Marlier, F., Sassetti, I., Blockaert, J., le Peuch, C., Sladeczek, F.A. The glutamate receptor of the Qp-type activates protein kinase C and is regulated by protein kinase C. *Neuroscience Letters,* 109; 146-51.

Marsala, M., Malmberg, A.B., Yaksh, T.L. The spinal loop dialysis catheter: characterization of use in the unanesthetized rat. *Journal of Neuroscience Methods,* 1995; 62:43-53.

Mayer, M.L., Miller, R.J. Excitatory amino acid receptors, second messengers and regulation of intracellular Ca2+ in mammalian neurons. In Lodge D, Collingridge GL (Eds.), *The Pharmacology of Excitatory Amino Acids, A TIPS Special Report* (pp. 36-42). Cambridge, M.A.: Elsevier.

McCall, W.D., Tanner, K.D., Levine, J.D. Formalin induces biphasic activity in c-fibers in the rat. *Neuroscience Letters,* 1996; 208:45-8.

McCarson, K.E., Goldstein, B.D. Release of substance P into the superficial dorsal horn following nociceptive activation of the hindpaw of the rat. *Brain Research,* 1991; 568:109-15.

McCarson, K.E., Krause, J.E. NK-1 and NK-3 type tachykinin receptor mRNA expression in the rat spinal cord dorsal horn is increased during adjuvant orformalin-induced nociception. *Journal of Neuroscience,* 1994; 14:712-20.

McCarson, K.E., Krause, J.E. The formalin-induced expression of tachykinin peptide and neurokinin receptor messenger RNAs in rat sensory ganglia and spinal cord is modulated by opiate preadministration. *Neuroscience,* 1995; 64:729-39.

Menendez, L., Perez-Vallina, J.R., Cantabrana, B., Hidalgo, A., Baamonde, A. Calmodulin inhibitors induce spinal analgesia in rats. *Brain Research,* 1996; 731:114-21.

Millan, M.J., Seguin, L. Chemically-diverse ligands at the glycine B site coupled to N-methyl-D-aspartate (NMDA) receptors selectively block the late phase of formalin-induced pain in mice. *Neuroscience Letters,* 1994; 178:139-43.

Millan, M.J., Seguin, L. (+)-HA 966, a partial agonist at the glycine site coupled to NMDA receptors, blocks formalin-induced pain in mice. *European Journal of Pharmacology,* 1993; 238:445-7.

Munglani, R., Jones, J.G., Hunt, S. Pre-emptive analgesia--Use of immediate early genes expression as markers of neuronal stimulation. *British Journal of Anaesthesia,* 1993; 71:458.

Murray, C.W., Cowan, A., Larson, A.A. Neurokinin and NMDA antagonists (but not a kainic acid antagonist) are antinociceptive in the mouse formalin model. *Pain,* 1991; 44:179-85.

Muth-Selbach, U.S., Tegeder, I., Brune, K., Geisslinger, G. Acetaminophen inhibits spinal prostaglandin E2 release after peripheral noxious stimulation. *Anesthesiology,* 1999; 91:231-9.

Nakanishi, O., Ishikawa, T., Imamura, Y. Modulation offormalin-evoked hyperalgesia by intrathecal N-type Ca channel and protein kinase C inhibitor in the rat. *Cellular and Molecular Neurobiology,* 1999; 19:191-7.

Naranjo, J.R., Mellström, B., Achaval, M., Sassone-Corsi, P. Molecular pathways of pain: Fos/Jun-mediated activation of a noncanonical AP-1 site in the prodynorphin gene. *Neuron,* 1991; 6:607-17.

Nässtrõm, J., Karlsson, U., Post, C. Antinociceptive actions of different classes of excitatory amino acid receptor antagonists in mice *European Journal Pharmacology,* 1992; 212 :21-9.

Nishiyama, T., Gyermek, L., Lee, C., Kawasaki-Yatsugi, S., Yamaguchi, T. The spinal antinociceptive effects of a novel competitive AMPA receptor antagonist, YM872, on thermal orformalin-induced pain in rats. *Anesthesia and Analgesia,* 1999a; 89:143-7.

Nishiyama, T., Gyermek, L., Lee, C., Kawasaki-Yatsugi, S., Yamaguchi, T. The systematically administered competitive AMPA receptor antagonist, YM872, has analgesic effects on thermal or formalin-induced pain in rats. *Anesthesia and Analgesia,* 1999b; 89:1534-7.

Nishizuka, Y. Studies and perspectives of protein kinase C. *Science,* 1986; 233:305-12.

Noguchi, K., Morita, Y., Kiyama, H., Ono, K., Tohyama, M. A noxious stimulus induces the preprotachykinin-A gene expression in the rat dorsal root ganglion: a quantitative study using in situ hybridization histochemistry. *Molecular Brain Research,* 1988; 4:31-5.

O'Connor, T.C., Abram, S.E. Inhibition of nociception-induced spinal sensitization by anesthetic agents. *Anesthesiology,* 1995; 82:259-66.

Ohkubo, T, Shibata, M., Takahashi, H.,Inoki, R.. Roles of substance P andsomatostatin on transmission of nociceptive information induced byformalin in spinal cord. *Journal of Pharmacology and Experimental Therapeutics,* 1990; 252:1261-8.

Oku, R., Nanayama, T., Satoh, M. Calcitonin gene-related peptide modulates calcium mobilization in synaptosomes of rat spinal dorsal horn. *Brain Research,* 1988; 475:356-60.

Puig, S., Sorkin, L.S. Formalin-evoked activity in identified primary afferent fibers -systemic lidocaine suppresses phase-2 activity. *Pain,* 1996; 64:345-55.

Rasmussen, H. The calcium messenger system. *New England Journal of Medicine,* 1986; 314:1094-1101.

Roche, A.K., Cook, M., Wilcox, G.L., Kajander, K.C. A nitric oxide synthesis inhibitor (L-NAME) reduces licking behavior and Fos-labeling in the spinal cord of rats during formalin-induced inflammation. *Pain,* 1996; 66:331-41.

Sakuruda, T., Katsumata, K., Manome, Y., Tan-No, K., Sakurada, S., Kisara, K., Ohba, M. Antinociceptive effects in the formalin and capsaicin tests after intrathecal administration of substance P analogues in mice. *European Journal of Pharmacology,* 1993a; 242:47-52.

Sakuruda, T., Katsumata, K., Yogo, H., Tan-No, K., Sakurada, S., Kisara, K. Antinociception induced by CP 96,345, a non-peptide NK-1 receptor antagonist, in the mouse formalin and capsaicin tests. *Neuroscience Letters,* 1993b; 151:142-5.

Sakuruda, T., Katsumata, K., Yogo, H., Tan-No, K., Sakurada, S., Ohba, M., Kisara, K. The neurokinin-1 receptor antagonist, sendide, exhibits antinociceptive activity in the formalin test. *Pain,* 1995; 60:175-80.

Sakuruda, T., Sugiyama, A., Sakurada, C., Tan-No, K., Yonezawa, A., Sakurada, S., Kisara, K. Effect of spinal nitric oxide inhibition on capsaicin-induced nociceptive response. *Life Sciences,* 1996; 59:921-30.

Seguin, L., Le Marouille-Girardon, S., Millan, M.J. Antinociceptive profiles of non-peptidergic neurokinin 1 and neurokinin 2 receptor antagonists: a comparison to other classes of antinociceptive agent. *Pain,* 1995; 61:325-43.

Shibuta, S., Mashimo, T., Zhang, P., Ohara, A, Yoshiya, I. A new nitric oxide donor, NOC-18, exhibits a nociceptive effect in the rat formalin model. *Journal of The Neurological Sciences,* 1996; 141:1-5.

Shimoyama, M., Shimoyama, N., Gorman, A.L., Elliott, K.J., Inturrisi, C.E. Oral ketamine is antinociceptive in the rat formalin test: role of the metabolite, norketamine. *Pain,* 1999; 81:85-93.

Shimoyama, M., Shimoyama, N., Elliott, K.J., Inturrisi, C.E. d-Methadone is antinociceptive in the rat formalin test *Journal of Pharmacology and Experimental Therapeutics,* 1997; 283:648-52.

Simmons, R.M., Li, D.L., Hoo, K.H., Deverill, M., Ornstein, P.L., Iyengar, S. Kainate GluR5 receptor subtype mediates the nociceptive response to formalin in the rat. *Neuropharmacology,* 1998; 37:25-36.

Skilling, S.R., Smullin, D.H., Beitz, A.J., Larson, A.A. Extracellular amino acid concentrations in the dorsal spinal cord of freely moving rats following veratridine and nociceptive stimulation. *Journal of Neurochemistry,* 1988; 51:127-32.

Sladeczek, F., Pin, J-P., Recasens, M., Bockaert, J., Weiss, S. Glutamate stimulates inositol phosphate formation in striatal neurons. *Nature,* 1985; 317:717-19.

Snyder, S.H. Nitric oxide: First in a new class of neurotransmitters. *Science,* 1992; 257:494-96.

Southam, E., East, S.J., Garthwaite, J. Excitatory amino acid receptors coupled to the nitric oxide: cyclic GMP pathway in rat cerebellum during development. *Journal of Neurochemistry,* 1991; 56:2072-81.

Sugiyama, H., Ito, I., Hirono, C. A new type of glutamate receptor linked to inositol phospholipid metabolism. *Nature,* 1987; 325:531-33.

Sun, W.Z., Shyu, B.C., Shieh, J.Y. Nitrous oxide or halothane, or both, fail to suppress c-fos expression in rat spinal cord dorsal horn neurones after subcutaneous formalin. *British Journal of Anaesthesia,* 1996; 76:99-105.

Takano, Y., Kuno, Y., Sato, E., Takano, M., Sato, I. The enhancement of formalin induced agitation behavior by intrathecal administration of prostaglandin E1. *Masui Japanese Journal of Anesthesiology,* 1999; 48:841-6.

Takano, Y., Sato, E., Kaneko, T., Sato, I. Antihyperalgesic effects of intrathecally administered magnesium sulfate in rats. *Pain,* 2000; 84:175-9.

Tao, Y.X., Hassan, A., Haddad, E., Johns, R.A. Expression and action of cyclic GMP-dependent protein kinase 1 alpha in inflammatory hyperalgesia in rat spinal cord. *Neuroscience,* 2000; 95:525-33.

Tao, Y.X., Zhao, Z.Q. NMDA receptors mediating Fos expression in rat spinal cord induced by subcutaneous injection of formalin. *Chung-Kuo Yao Li Hsueh Pao - Acta Pharmacologica Sinica,* 1998; 19:506-9.

Tao, Y.X., Zhao, Z.Q. Ultrastructure of Fos-labeled neurons relating to nociceptive primary afferent and substance P terminals in rat spinal superficial laminae. *Neuropeptides,* 1997; 31:327-32.

Taylor, B.K., Peterson, M.A., Basbaum, A.I. Persistent cardiovascular and behavioral nociceptive responses to subcutaneous formalin require peripheral nerve input. *Journal of Neuroscience,* 1995; 15:7575-84.

Taylor, B.K., Peterson, M.A., Basbaum, A.I. Early nociceptive events influence the temporal profile, but not the magnitude, of the tonic response to subcutaneousformalin: effects with remifentanil. *Journal of Pharmacology and Experimental Therapeutics,* 1997; 280:876-83.

Tokunaga, A., Doi, M., Senba, E. Effects of local anaesthesia on formalin-induced fos expression in the rat dorsal horn. *Neuroreport,* 1995; 6:2301-4.

Traub, R.J. The spinal conribution of substance P to the generation and maintenance of inflammatory hyperalgesia in the rat. *Pain,* 1996; 67:151-61.

Vaccarino, A.L., Marek, P., Kest, B., Weber, E., Keana, J.F.W., Liebeskind, J.C. NMDA receptor antagonists, MK-801 and ACEA-1011, prevent the development of tonic pain following subcutaneous formalin. *Brain Research,* 1993; 615:331-4.

Wang, Y.X., Gao, D., Pettus, M., Phillips, C., Bowersox, S.S. Interactions of intrathecally administered ziconotide, a selective blocker of neuronal N-type voltage-sensitive calcium channels, with morphine on nociception in rats. *Pain,* 2000; 84:271-81.

Wheeler-Aceto, H., Cowan, A. Neurogenic and tissue-mediated components offormalin-induced edema: Evidence for supraspinal regulation. *Agents Actions,* 1991; 34:264-9.

Womack, M.D., MacDermott, A.B., Jessell, T.M. Sensory transmitters regulate intracellular calcium in dorsal horn neurons. *Nature,* 1988; 334:351-53.

Womack, M.D., MacDermott, A.B., Jessell, T.M. Substance P increases [Ca2+i in dorsal horn neurons via two distinct mechanisms. *Society for Neuroscience Abstracts,* 1989; 15:184.

Woolf, C.J., Chong, M-S. Preemptive analgesia -- treating postoperative pain by preventing the establishment of central sensitization. *Anesthesia and Analgesia,* 1993; 77:362-79.

Yamamoto, T., Nozaki-Taguchi, N. The effects of intrathecally administered FK480, a cholecystokinin-A receptor antagonist, and YM022, a cholecystokinin-B receptor antagonist, on the formalin test in the rat. *Anesthesia and Analgesia,* 1996; 83: 107-13.

Yamamoto, T., Shimoyama, N., Mizuguchi, T. Effects of intrathecal FK888, a novel dipeptide NK1 receptor antagonist, on the formalin test in the rat. *Neuroscience Letters,* 1993a; 161:57-9.

Yamamoto, T., Shimoyama, N., Mizuguchi, T. Nitric oxide synthase inhibitor blocks spinal sensitization induced by formalin injection into the rat paw. *Anesthesia and Analgesia,* 1993b; 77:886-90.

Yamamoto, T., Yaksh, T.L. Comparison of the antinociceptive effects of pre- and posttreatment with intrathecal morphine and MK801, an NMDA antagonist, on the formalin test in the rat. *Anesthesiology,* 1992; 77:757-63.

Yamamoto, T., Yaksh, T.L. Stereospecific effects of a nonpeptidic NK1 selective antagonist, CP-96,345: Antinociception in the absence of motor dysfunction. *Life Sciences,* 1991; 49:1955-63.

Yashpal, K., Pitcher, G.M., Parent, A., Quirion, R., Coderre, T.J. Noxious thermal and chemical stimulation induce increases in 3H-phorbol 12,13-dibutyrate binding in spinal cord dorsal horn as well as persistent pain and hyperalgesia, which is reduced by inhibition of protein kinase C. *Journal of Neuroscience,* 1995; 15:3263-72.

Yashpal, K., Katz, J., Coderre, T.J. Effects of preemptive or postinjury intrathecal local anesthesia on persistent nociceptive responses in rats: Confounding influences of peripheral inflammation and the general analgesic regimen. *Anesthesiology,* 1996; 84:1119-28.

Yashpal, K., Mason, P., McKenna, J.E., Sharma, S.K., Henry, J.L., Coderre, T.J. Comparison of the effects of treatment with intrathecal lidocaine before and after formalin on both nociception and Fos expression in the spinal cord dorsal given horn. *Anesthesiology,* 1998; 88:157-64.

Yashpal, K., Radhakrishnan, V., Coderre, T.J., Henry, J.L. CP-96,345, but not its stereoisomer, CP-96,344, blocks the nociceptive responses to intrathecally administered substance P and to noxious thermal and chemical stimuli in the rat. *Neuroscience,* 1993; 52:1039-47.

Zhang, R.X., Mi, Z.P., Qiao, J.T. Changes of spinal substance P, Calcitonin gene-related peptide, somatostatin, Met-enkephalin and neurotensin in rats in response to formalin-induced pain. *Regulatory Peptides,* 1994; 51:25-32.

8 NEURAL DARWINISM IN THE MAMMALIAN SPINAL CORD

V. Reggie Edgerton
Brain Research Institute
Department of Physiological Science
Department of Neurobiology
University of California
Los Angeles, CA 90095

Roland R. Roy
Brain Research Institute
University of California
Los Angeles, CA 90095

Ray D. de Leon
Brain Research Institute
University of California
Los Angeles, CA 90095

8.1 INTRODUCTION

A conceptual framework around which the spinal control of locomotion is viewed and often studied has focused on the property of "Central Pattern Generation", i.e.

the generation of oscillating efferent patterns from the lumbosacral spinal cord in the absence of either supraspinal or sensory oscillating input (Grillner, 1973; 1982; Severston, 1999). An alternative, or at least complementary, concept that may help to understand spinally-controlled locomotion has been presented by Edelman and colleagues (Edelman, 1978; 1981; Edelman and Reeke, 1982; Edelman and Finkel, 1984) to account for adaptive events in the brain. They theorized that "the brain is dynamically organized into cellular populations containing individually varied networks, the structure and function of which are selected by different means during development and behavior". The cellular populations are proposed to be collections of hundreds to thousands of strongly interconnected neurons acting as functional units. The key elements of this Neuronal Group Selection concept, often referred to as Neural Darwinism, are a) "Reentrant" structure and function involving a continuous spatial temporal representation of sensory information and a mechanism for continuous updating of the selection of neuronal groups that can generate a given motor synergy or pattern and, b) "Degeneracy" indicating that there must be variance in the neuronal groups that generate a movement pattern if learning and selective matching of sensory and motor processes are to occur. This concept, therefore, also assumes that different neuronal groups can accomplish the same function. It is further proposed that the adaptive nature of these functional units reflect neural projections formed either during development (primary repertoire) or as a result of experience and changing synaptic efficacy (secondary repertoire). Clear examples of the same neuron contributing to multiple pattern-generating circuits in the stomatogastric ganglion have been described (Weimann and Marder, 1994).

Although the concepts of Neuronal Group Selection and Central Pattern Generation are not contradictory, the functional features which each emphasizes differ substantially. The former concept emphasizes the role of sensory input, whereas the latter concept inherently focuses on the intrinsic network features that generate repetitive step cycles in the case of locomotion in the absence of sensory information. Ironically, the characteristic of Central Pattern Generation that has been emphasized experimentally, i.e. oscillatory efferent output without oscillating input from supraspinal or peripheral afferents, may not represent its most important feature. We suggest that the feature of greater importance is the ability to interpret afferent information and generate motor responses appropriate for a range of conditions recognized by ensembles (cellular populations) of afferent information. The results from a series of experiments that we have performed over the last 20 years are consistent with both models playing important roles in the recovery of locomotion following spinal cord injury. Although the Neuronal Group Selection theory generally has been assumed to be relevant only for supraspinal elements, i.e. the brain, our data suggest that this theory is equally relevant to neural pathways in the spinal cord.

8.2 INTEGRATION OF EVENTS RANGING FROM THE RELEASE OF SYNAPTIC VESICLES TO LOCOMOTION: A PROBABILITY PERSPECTIVE

A number of extensive reviews have detailed the experiments demonstrating that after a complete low-thoracic spinal cord transection adult cats can generate full weight-bearing stepping over a range of speeds and at varying loads that is kinematically and kinetically similar to, but not the same as, that observed in the intact cat (Duysens, Clarac and Cruse, 2000; Edgerton, de Guzman, Gregor, Roy, Hodgson and Lovely, 1991; Edgerton, Roy, Hodgson, Prober, de Guzman, and de Leon, 1991; Edgerton, Roy, Hodgson, Prober, de Guzman and de Leon, 1991; Edgerton, de Leon, Tillakaratne, Recktenwald, Hodgson and Roy, 1997a; 1997b; Grillner, 1973; Orlovsky, Deliagina and Grillner, 1999; Rossignol, 1996). These spinal animals also can execute prolonged standing (de Leon, Hodgson, Roy and Edgerton, 1998b; Pratt, Fung and Macpherson, 1994), paw shaking (Barbeau and Rossignol, 1990; Macpherson and Fung, 1999) and can continue stepping when a stumbling stimulus is introduced to the hindlimb during the swing phase of a step (Forssberg, Grillner and Rossignol, 1977; Nakada, Hodgson, de Leon, Roy and Edgerton, 1994; Rossignol and Gauthier, 1980). Thus, it is clear that there are selected and predictable combinations of motoneurons that are activated for a given motor response when a particular pattern of sensory information is projected to the lumbosacral spinal cord of animals in which all supraspinal input has been eliminated. The successful generation of these motor responses in complete spinal animals is highly dependent on the temporal patterns of the afferent information associated with the execution of the specific task.

Several general principles seem to apply to spinally-controlled motor behaviors such as stepping. First, the level of success in the execution of a motor task reflects the cumulative probability of a sequence of events occurring at any given time (Fig. 8.1). Second, these behaviors reflect probabilistic events that occur at multiple levels of integration (Fig. 8.1). These probabilistic events are relevant to whether or not there is release of neurotransmitters from a single vesicle at a synapse to the release of vesicles associated with a population of synapses at a single neuron. These events determine the changes in the membrane potential of a single neuron and combinations of neurons that represent an ensemble of excitatory and inhibitory events that excite a given combination of motor pools and a given number of motoneurons within those motor pools, at any given instant in time. Third, there are probabilities associated with the determination of which combination of ensembles of synaptic events and neural pathways that can be excited at any given time or phase of a successful motor task such as stepping. Even in the normally intact motor system, i.e. with supraspinal control present, the notion of a hard-wired control system in the mammalian spinal cord is inconsistent with experimental evidence. For example, there are significant variations in the kinematics and kinetics of the motor patterns across steps even on a treadmill under highly controlled and constant conditions in an intact, uninjured animal. Fourth, the probability of events is modulated continuously over multiple time frames, i.e. milliseconds, seconds, minutes, hours, days or months, each most likely occurring through different mechanisms. Fifth, the probability of a given set of sensorimotor pathways being active at any time is a function of those pathways which were activated prior to that time point, i.e. each combination of events impacts those events which are likely to occur next. Ensembles of neurons that have the intrinsic properties to generate complex, oscillatory motor outputs with a high level of predictability can simulate locomotion without supraspinal or afferent oscillatory input, i.e. Central Pattern Generation (Pearson, 1995; Prochazka and Gorassini, 1998a; 1998b). These five general principles of the locomotor apparatus apply equally well following low-thoracic spinal cord transection, although significant synaptic reorganization occurs following spinal cord injury. Many of the details of the reorganization seem to be a function of the sensorimotor experience of the spinal cord. These adaptive properties suggest that the neural networks in the spinal cord can dynamically reorganize into different cell populations continuously. These pathways do not function as a hard-wired system.

Vesicular Neurotransmitter Release

→ Synapse Modulation

→ Neuron Action Potential

→ Interneuronal Populations

→ Populations of Motor Units

→ Populations of Motoneurons

Figure 8.1. Increasing levels of organization of functional components, ranging from vesicular neurotransmitter release to populations of motoneurons.

As noted previously, the probability for a successful step cycle is an integrated function at multiple levels of organization (Fig. 8.1), i.e. from a synapse to the complex sequences of functional pathways activated in all muscles that contribute to a step cycle. For example, some sequence of theoretical events, i.e. A→B→C→D→E, may successfully generate standing or stepping. However, there may be more than one theoretical sequence and combination of the sequences that can generate these motor tasks. The sequencing of all of the potential neural pathways that contribute to locomotion may occur randomly or perhaps non-randomly, but a step will be successful only when there is some critical level of temporal and spatial order of excitatory and inhibitory events, i.e. highly non-random. The probability of executing consecutive successful steps adds another increment of difficulty, or reduced likelihood of success, but remains a function of the probability of executing a single successful step. For example, the probability of a complete spinal cat executing ten or more consecutive steps successfully after being trained for 6 weeks to step on a treadmill (5 days per week, 30 min/day) was 11%, whereas a spinal cat that was not trained was unable to generate ten consecutive steps at any time, i.e. 0% (Fig. 8.2).

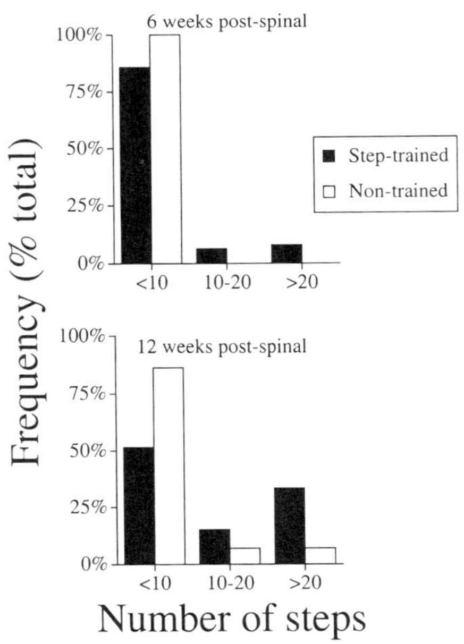

Figure 8.2. The incidence (% total) of non-trained and step-trained spinal cats that could perform less than 10, 10-20 or more than 20 consecutive full weight-bearing steps on a treadmill at 0.4 meters/sec., 6 or 12 weeks post-spinalization. There was a significant relationship between training state and stepping ability at the 12, but not the 6, week time period, i.e. Chi square: $p = 0.001$ and < 0.08, respectively.

After 12 weeks the trained spinal cats executed ten or more consecutive successful steps in ~35% of the trials. At this same stage, the non-trained spinal cats were successful in ~10% of the trials. These results show that some level of recovery from spinal transection occurs spontaneously, i.e. attributable to the reestablishment of some functional neural connections which are mediated by "healing" mechanisms intrinsic to the spinal cord below the level of the lesion. But even these spontaneous mechanisms can be influenced by the routine cage activity experienced throughout the day, thus suggesting that even spontaneous recovery is, in part, activity-dependent. However, the enhanced probability of successfully executing longer bouts of consecutive stepping as a result of only 30 min of step training per day demonstrates a robust activity-dependent plasticity that reflects learning a motor task. Similar results were obtained when we compared the ability of cats that were trained to stand vs. those that received no stand training (Fig. 8.3).

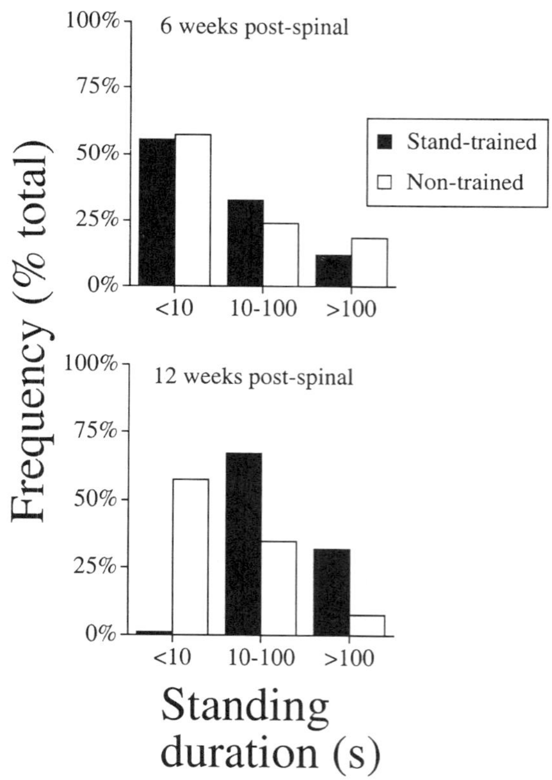

Figure 8.3. The incidence (% total) of non-trained and stand-trained spinal cats that could perform less than 10, 10-20 or more than 20 seconds of continuous full weight-bearing standing 6 or 12 weeks post-spinalization. There was a significant relationship between training state and standing ability at the 12, but not the 6, week time period, i.e. Chi square: $p = 0.001$ and < 0.32, respectively.

In this case, the probabilities are expressed as the likelihood of a cat maintaining continuous standing, without perturbation or assistance by the experimenter, for different durations. Comparisons of the probabilities of executing a motor task successfully before and after training spinal cats to step or stand illustrate that the degree of randomness of the sensorimotor events that execute a motor task can be modulated by the sensorimotor "experience" of the spinal cord (de Leon, Hodgson, Roy and Edgerton, 1998b; 1999a; Durkovic, 1986; Hodgson, Roy, de Leon, Dobkin and Edgerton, 1994; Lovely, Gregor, Roy and Edgerton, 1986; 1990; Roy, Baldwin and Edgerton, 1991; Roy, Talmadge, Hodgson, Zhong, Baldwin and Edgerton, 1998).

The changing probabilities of the relative levels of activation of two motor pools as a result of spinal cord transection followed by step training are illustrated in Figure 8.4.

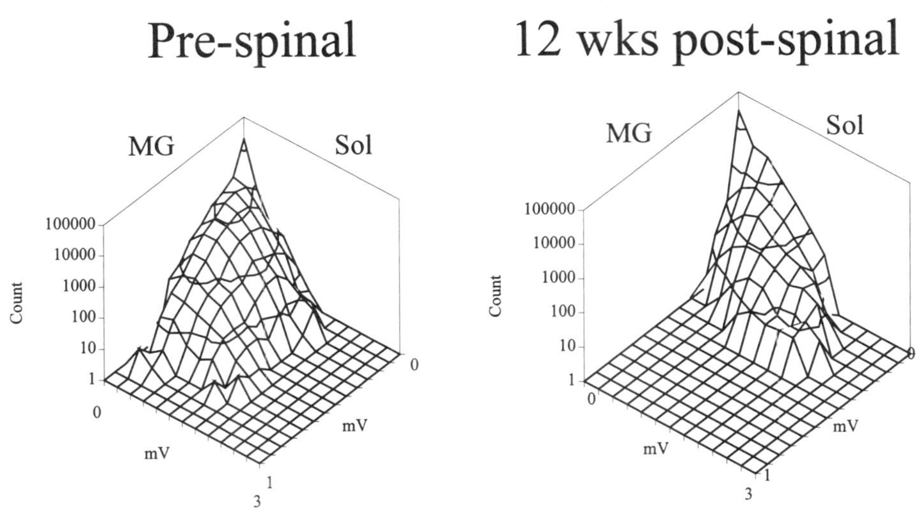

Figure 8.4. EMG amplitude joint probability density distributions for the soleus (Sol) and medial gastrocnemius (MG) muscles during 40 cycles of stepping on a treadmill at 0.4 meters/sec. pre-spinalization or after 12 weeks of step training post-spinalization. The range of EMG amplitudes was 0 to 3 mV for the Sol and 0 to 1 mV for the MG.

The incidence of different pairs of EMG amplitudes recorded from the soleus and the medial gastrocnemius muscles over a series of weight-bearing steps differs in the pre-spinal and the 12-week post-spinal trained condition. After 12 weeks of step training, the EMG amplitudes showed a greater bias toward soleus than medial gastrocnemius activation. In other words, the higher EMG amplitudes occurred more frequently in the soleus than the medial gastrocnemius. These data also are consistent with the observation that there are multiple combinations of "events" or neural pathways that can execute the same motor task. This concept can be extended to the recovery of stepping in spinalized cats, i.e. hindlimb locomotion in spinal cord injured animals, even when they are trained to step, differs from that of intact animals. It remains to be determined, however, at what level of integration these new adaptive strategies occur that enable the spinal animal to regain the ability to generate load-bearing stepping.

8.3 MULTIPLE SOLUTIONS TO LOCOMOTION

There does not appear to be a "fixed" set of neural pathways activated to generate each step in a successful series of step cycles. This is reflected in the fact that even under the most controlled conditions each step has its unique kinematic and kinetic features. At the same time, there is some limit or "envelope" of flexibility that the spinal cord can tolerate while continuing to execute and sustain "successful" stepping. Examples of the variability in activation patterns of the soleus and medial gastrocnemius muscles within a single step and across two consecutive steps are shown in Figure 8.5.

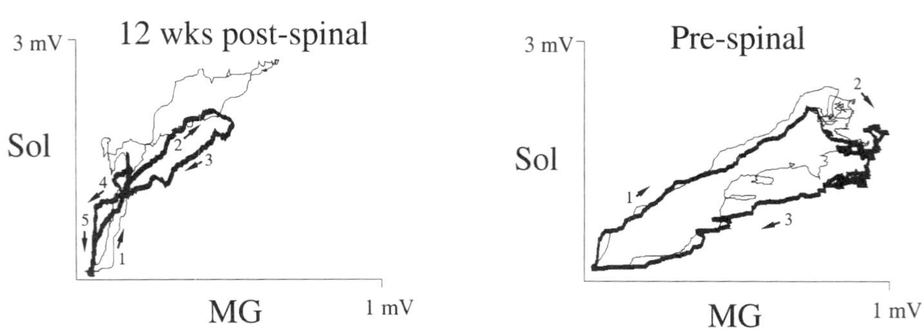

Figure 8.5. Scatterplots for the EMG activity of the soleus (Sol) and medial gastrocnemius (MG) muscles during stepping on a treadmill at 0.4 meters/sec. pre-spinalization or after 12 weeks of step training post-spinalization. Scatterplots for two steps are shown (thick and thin lines). The numbers with arrows indicate the progression of the scatterplots for the step shown as a thick line).

In this scatterplot the sequences of the relative EMG amplitudes for each muscle are plotted. Variations in the details of these sequences of EMG amplitude as a step progresses are evident. Also note that the variation differs between the two consecutive steps, both of which were equally successful. These data further illustrate multiple solutions to the same motor task. Figure 8.4 demonstrates the changing probability of the relative activation of the soleus and medial gastrocnemius muscles after 12 weeks of training following spinalization compared to stepping in the same animal before spinalization. The differences between the pre-spinal (with an intact nervous system) and post-spinal (with the neural pattern attributable to the spinal networks) patterns again demonstrate that there are multiple solutions to executing successful stepping. To understand the fundamentals of the neural control used to accommodate multiple solutions within a series of steps even during normal locomotion, a probabilistic perspective seems applicable and useful. This probabilistic perspective seems even more essential in understanding how the neural pathways in the spinal cord can control stepping following spinal cord injury.

How does a probabilistic approach conceptually or experimentally help to understand the neuromotor control of the spinal cord? Can this approach be used to assess the changing properties of each of the elements of stepping? For successful locomotion, there is a limited number of neurons from a limited combination of motor pools and a given number of motoneurons in each motor pool that can be activated at any given point during each step. Thus, a successful step requires that some limited number, kinds, and "spatially" appropriate pathways be activated within an acceptable temporal envelope (de Guzman, Roy, Hodgson and Edgerton, 1991). In the generation of a successful step, the probability of pathways A, B, C, D, and then E, or some functionally equivalent pathways, being activated in the appropriate sequence will be high (Fig. 8.1).

8.4 CENTRAL PATTERN GENERATION AND LOCOMOTION

The ability to activate or "bias" selected neuronal pathways towards excitation or inhibition within a limited time frame and sequence may be one of the more important features of what is commonly recognized as Central Pattern Generation (Baev and Konstantin, 1998; Severston, 1999; Stein, 1999). In effect, the properties of locomotor Central Pattern Generation are manifestations of a progression of changing, but predictable, events throughout all phases of a "step cycle". This predictability does not reflect one unique combination of neural pathways that will be activated to accomplish the task. In normal animals, the brain with all of its sensory input is intact and undoubtedly contributes to the “noise” in stepping, but at the same time these inputs during Central Pattern Generation provide additional sources of control and, in effect, provide a system that is more “tolerant” to variations because it can be accommodated. The multiple combinations of activation patterns that occur within some envelope of acceptable variance is reflected in the variation in EMG and kinematics that is normally observed from step to step (Fig. 8.5).

A probabilistic approach also can be applied to a range of neurophysiological phenomena such as acclimation, habituation, post-tetanic potentiation, "warm-up," and different forms of learning. As noted previously, the probabilistic concept can be applied to a range of time frames as well, i.e. from milliseconds to seconds or even minutes, days, weeks or months. The results from some experimental models in which spinal cord "learning" lasts over a period of months may be explained by the changing probabilistic properties associated with functionally reorganizing cellular populations as opposed to the growth or elimination of selected neural pathways. This interpretation contrasts with the view that changes in anatomically-defined neural connections underlie the mechanism(s) by which neural plasticity occurs. Based on this anatomical perspective, some have concluded erroneously that neural pathways in the spinal cord have little plasticity. Certainly the changing physiological-behavioral features of the spinal sensorimotor systems that generate locomotion need not occur as a result of the reconstruction of neural pathways in a “Cajal sense”.

8.5 SENSORY PROCESSING BY THE SPINAL CORD

Although Central Pattern Generators can generate locomotor-like patterns without any oscillating central or peripheral input, this does not imply a lesser, or even nonessential, role for afferents during normal or spinal locomotion. It is true that an acutely spinalized and deafferented cat can generate oscillatory movements of the hindlimbs on a treadmill when pharmacologically stimulated with drugs, e.g., clonidine (Barbeau, Julien and Rossignol, 1987; Barbeau and Rossignol, 1990; 1991; Grillner, 1981; Rossignol, Barbeau and Julien, 1986; Rossignol and Barbeau, 1993). However, this locomotion is not full weight bearing and the rate of stepping can only accommodate the treadmill belt speed which matches the intrinsic oscillatory rate of the efferent output. Although complete spinal animals can regain full weight-bearing stepping, the dorsal lumbosacral afferent roots must be intact. Chronic spinal animals are unable to regain any effective weight-bearing stepping if the dorsal roots are severed, i.e. the spinal cord isolation model (Roy, Hodgson, Lauretz, Peirotti, Gayek and Edgerton, 1992; Tower, 1937). Chronic EMG recordings have confirmed that after spinal cord isolation, the motor pools are electrically silent and the limbs remain flaccid and this silence can be maintained for months or even years (Pierotti, Roy, Bodine-Fowler, Hodgson and Edgerton, 1991; Roy, Eldridge, Baldwin and Edgerton, 1996). As is the case after acute spinalization plus deafferentation, however, the motor pools can be activated pharmacologically in an oscillatory pattern months after a low thoracic, complete spinal cord transection (Barbeau, Julien and Rossignol, 1987; Barbeau and Rissignol, 1991; Chau, Barbeau and Rossignol, 1998a; 1998b; de Leon, Tamaki, Hodgson, Roy and Edgerton, 1999b; Robinson and Goldberger, 1986; Rossignol, Barbeau and Julien, 1986; Rossignol and Barbeau, 1993; Stewart, Barbeau and

Gauthier, 1991). Thus, afferent input to the spinal cord is an essential component for the effective generation of stepping, i.e. full weight-bearing stepping that can accommodate variable treadmill belt speeds at varying loads. Given these observations, a critical property of the spinal Central Pattern Generator is the ability to interpret and respond to the afferent input from the hindquarters with a level of preciseness and consistency that will enable the control system to accommodate to the changing physical environment.

8.6 PHASE DEPENDENT MODULATION OF AFFERENT INPUT TO THE SPINAL CORD

Further evidence for the decision-making ability of the spinal cord is the observation that polysynaptic cutaneous input affects the motor output of most, if not all, muscles of both hindlimbs during locomotion in the chronic spinal animal. Whether flexor or extensor motor pools of a given limb are activated or inhibited is a function of the timing of the cutaneous stimulus relative to the phases of the step cycle. For example, the spinal cord receives and interprets the cutaneous information from the dorsum of one paw as a function of the specific phase of the step cycle in which it occurs (Forssberg, Grillner and Rossignol, 1997; Rossignol and Gauthier, 1980). The response is not constant in amplitude nor in polarity during all phases of a step. For example, when the dorsum of the foot is stimulated during the swing phase of the step cycle, the flexor muscles of the ipsilateral leg are activated whereas the extensor muscles are reciprocally inhibited. In the contralateral limb, the extensors are activated whereas the flexor motor pools are inhibited. A similar stimulus applied during the stance rather that the swing phase of the same limb induces the opposite response, i.e. excitation of extensors ipsilaterally and flexors contralaterally. Thus, an "interpretive" function is clearly intrinsic to these spinal pathways, supporting the view that a critical feature of the locomotor Central Pattern Generator is the interpretation of afferent data in a functional context. Similar phase-dependent modulation of the H-reflex in humans has been demonstrated, although the patterns of modulation differ somewhat from those observed in cats (Capaday and Stein, 1987).

We extended the phase-dependent modulation of the stumbling reflex from the intra-step time frame to a more prolonged period, i.e. reflex modulations that persisted over multiple steps (Edgerton and Roy, 1994; Nakada, Hodgson, de Leon, Roy and Edgerton, 1994). When we induced the stumbling response in a series of 5-50 consecutive steps, the paw of a complete spinal cat was elevated to a higher level during the swing phase of the steps after, than before, the repeated stumbling stimuli were induced (Fig. 8.6).

In Figure 8.6, the dashed line from steps 0 to 10 represents a consistent swing height prior to the initial stumbling stimulus occurring at step 10. Note that the maximum height of the paw is elevated throughout the next 9 steps when the stumbling stimulus continues to be present during the swing phase of each step.

Upon removal of the stumbling stimulus at step 20, the height of the swing is reduced, but remains above the pre-stumbling step series for the next 10 steps (steps 21-30). In most cases, the height of the paw remained elevated for 2-5 steps immediately after the stimulus was removed. This graph illustrates an adaptive phenomenon that occurs within the time frame of a single step and shows evidence of a "memory" process dissipating over a period of several steps as well as one lasting over perhaps minutes as indicated by the sustained elevated swing at the end of the stepping series.

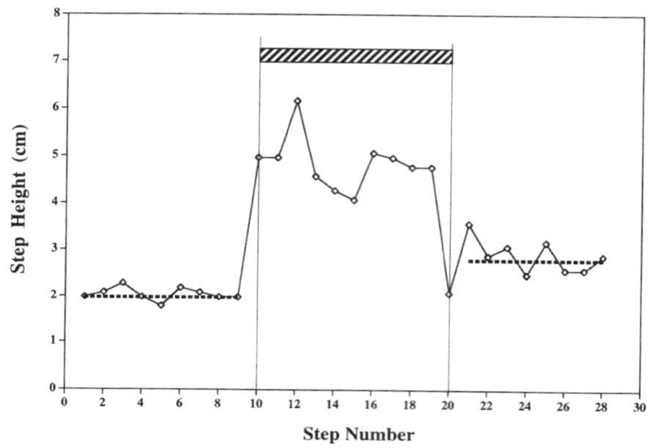

Figure 8.6. Relationship between step height and step number in a spinal cat during 29 consecutive steps on a treadmill at 0.4 meters/sec. A bar (stumbling stimulus) was placed in the path of the trajectory of the limb during swing from step 10 to step 19. The dashed lines represent the mean step height for the steps taken before (steps 1 to 9) and after (steps 20 to 29) the stumbling stimulus was present.

Several important points have been demonstrated in previously reported experiments as well as in the present data shown in Figure 8.6 involving the stumbling response:

1) The motor response to afferent input is a function of the phase of the step cycle (Forssberg, Grillner and Rossignol, 1997; Rossignol and Gauthier, 1980), thus showing an important context-dependent decision making process within the spinal cord;
2) The phase-dependent responses reflect a progressive modulation of the probability of excitation of a given set of motor pathways (Forssberg, Grillner and Rossignol, 1997; Rossignol and Gauthier, 1980); and
3) The phase dependence of the response suggests that the spinal cord responds to an ensemble of proprioceptive-cutaneous input that informs the spinal networks of the current phase of the step cycle thus enabling it to "anticipate" or "predict" the activation patterns appropriate to complete the next phase of the cycle (Hodgson, Roy, de Leon, Dobkin and Edgerton, 1994 and Fig. 8.6).

This phase- or state-dependent modulation is a central and critical feature of the spinal neuromotor networks. This "smartness" may be attributable to the same pathways that can generate oscillating patterns of output, i.e. the Central Pattern Generators. From an evolutionary perspective the ability of the spinal cord to execute complex postural and locomotor tasks with intact proprioceptive inputs but no supraspinal processing might be expected. To have some intrinsic automatism within the spinal cord has considerable advantages given that the appropriate motor responses in a 1G environment to maintain an upright posture and to sustain locomotion have a limited number of options (1). Likewise there will be a limited number of combinations of ensembles of afferent information, including the temporal features of the progression of the changing ensemble. The evolution of the sensorimotor pathways within the spinal cord of animals in Earth's gravitational environment obviously has accommodated the demands to successfully control many of the neural control details for a wide range of motor tasks at the spinal level. Intuitively, it seems that the more information processing that can occur at the spinal level the freer the brain can be to execute motor tasks in less predictable paradigms than the mechanics needed to step or stand in a known gravitational environment.

8.7 OTHER EXAMPLES OF MODULATION OF THE PROBABILITIES OF MOTOR EVENTS

Spasticity is a term that often raises immediate controversy and confusion. In the present chapter, we use the term to refer to contractions that occur spontaneously or in response to some sensory cue, e.g., pinching or heating the skin or suddenly stretching a muscle. The resulting contractions are involuntary and usually non-useful. The primary point relevant to this chapter is that spasticity has a low predictability and, therefore, level of useful control in modulating a given set of pathways in a desired sequence as is necessary to perform a motor task consistently. In addition to spastic contractions occurring randomly, the contractions of multiple muscles generally do not occur in a sequence that can generate useful movements. On the other hand, the presence of spasticity in a spinal cord injured subject demonstrates that the same motor pools have the potential to generate neuromuscular activity that can contribute to stepping. This activation occurs, however, in inappropriate patterns and with inappropriate combinations of motor pools. The process of training a spinal cat to step or stand is essentially increasing the probability of a given combination of motor pools being activated in a pattern that will result in successful stepping or standing. To change spasticity to effective locomotion requires a functional reorganization of pathways that will activate the muscles in sequences and amplitudes that generate effective movements. The challenge is to facilitate the development of the required sequences and levels of activation. Repetitive performance of a motor task, such as stepping and standing, can apparently improve the level of coordination among the relevant pathways that generate these tasks effectively in the complete spinal cat (de Leon, Hodgson, Roy and Edgerton, 1998a; 1998b; 1999a). It appears that a similar phenomenon occurs in humans with incomplete spinal cord injuries (Wernig, Muller, Nanassy and Cagol, 1995). Although, human subjects with complete spinal cord injuries have the capability of organizing the temporal patterns and amplitude of activation appropriate for stepping with training, the level of plasticity obtained to date has not been sufficient to generate independent stepping (Dietz, Colombo, Jensen and Baumgartner, 1995; Harkema, Hurley, Patel and Requejo, 1997; Wernig, Muller, Nanassy and Cagol, 1995).

8.8 PHARMACOLOGICAL MODULATION OF THE PROBABILITIES OF NEURAL PATHWAY ACTIVATION

The probability of activating the appropriate pathways in the temporal and spatial patterns necessary for stepping can be modulated pharmacologically as well as by repetitive training. A common pharmacological intervention for spinal cord injured subjects has been baclofen, which acts as an agonist for $GABA_b$ receptors and reduces spasticity in some spinal cord injured subjects. The net effect of using this particular intervention is to reduce the net excitability of the spinal cord and, therefore, reduce muscle spasticity. Its effectiveness varies among subjects and within a subject over time. Although this drug treatment may minimize the occurrence of spasticity, this medication in most spinal cord injured subjects theoretically would reduce the probability of successfully generating stepping by limiting afferent processing.

A concept that must be recognized in the pharmacological management of spinal cord injured subjects is the highly state-dependent properties of the injured spinal cord (Edgerton, de Leon, Tillakaratne, Recktenwald, Hodgson and Roy, 1997a). For example, a dose of strychnine, which selectively blocks glycine-mediated inhibition, has no detectable effect on the stepping performance of a normal cat and a barely detectable, although positive, effect on the stepping ability of a complete spinal cat that has been trained to step over a period of weeks (de Leon, Tamaki, Hodgson, Roy and Edgerton, 1999b). In contrast, the same dosage administered to a complete spinal cat that has been trained to stand and unable to generate weight-supporting stepping enables successful full weight-bearing stepping over a range of speeds within 30 min of administration. These findings can be interpreted in the context of the probability of pathway excitability as described above. The reduction in the level of inhibition by strychnine appears to be very selective since the coordination, (the temporal sequence of the activation of agonistic and antagonistic motor pools of both hindlimbs) and the amplitude of activation within the relevant motor pools are sufficient to generate stepping (de Leon, Tamaki, Hodgson, Roy and Edgerton, 1999b). In all reports of pharmacological modulation of the locomotor capacity of spinal animals, however, there has been marked animal to animal variability for every neurotransmitter-modulator system that has been studied (Rossignol, Barbeau and Julien, 1986; Rossignol and Barbeau, 1993). In many cases, the probabilistic perspective of "state dependence" after spinal cord injury can account for much of the variability in the responses. The general principle seems to follow that the change in probabilities of excitation imposed pharmacologically on the spinal sensorimotor system will depend on the background of probabilities present at the time of administration, i.e. state dependent.

The stychnine-induced modulation of the probabilities involved in the generation of stepping is interesting in another context. For example, a continuous and monotonic level of inhibition in all motor pools throughout a step cycle would have no direct beneficial effect on the capacity to step. That is, the presence of strychnine itself does not directly induce oscillating levels of inhibition in a temporally appropriate sequence for stepping. Neither does strychnine induce oscillation, i.e. pattern generation, in the adult spinal cord in the absence of oscillatory-afferent input. The ability of strychnine to induce stepping *in vivo* following complete spinal cord transection and stand training seems to be due to a change in the probability of excitation that is generated by the afferent input that is associated with the kinetics and kinematics of weight-bearing stepping. The "state-dependence" phenomenon is demonstrated by the observation that the spinal cats that could not step because they were trained to stand, could step within a short period after strychnine administration. Stepping in spinal animals that could already step because they had already been trained to step for 12 weeks showed little change in their kinematics and EMG patterns in response to the same strychnine dose that induced stepping in the stand-trained spinal cats. Thus the changes in probabilities associated with strychnine-induced stepping are a function of the probability states associated with stepping without the drug.

8.9 EVIDENCE OF NEURAL DARWINISM IN THE HUMAN SPINAL CORD

In some of our initial studies of spinal cord injured humans we hypothesized that the human spinal cord could process the sensory information derived from the lower limbs during weight-bearing stepping. We further hypothesized that the spinal cord could process this information in a manner that could be used to generate step-like dynamics. A fundamental component of these hypotheses was that the spinal cord could interpret all of the information derived from the lower limbs during stepping as a continuously changing ensemble of sensory information from which its relevance could be interpreted and utilized in a phase-dependent manner. In essence, this means that the spinal cord is "aware" of some combination of the mode, source and temporal pattern of afferent input associated with load-bearing stepping.

To study these possibilities, we mechanically assisted the stepping motions of subjects with a complete spinal cord injury while suspended above a treadmill in a parachute-like vest as described previously (Harkema, Hurley, Patel and Requejo, 1997). These experiments demonstrated that the human spinal cord could interpret the sensory information associated with the level of loading such that the EMG amplitudes of muscles controlling the ankle were modulated in proportion to the level of loading (Fig. 8.7).

In addition to modulating the level of EMG amplitudes in flexor and extensor muscles in proportion to the level of loading on the limbs, the modulation in EMG amplitude occurred in a phase-dependent manner. This phase dependence was evident over a range of speeds of stepping and loading. In addition we demonstrated that the modulation in the EMG patterns could not be accounted for solely by any stretch reflex mechanism of any one muscle or group of synergistic muscles. Although the results were consistent with stretch having some influence, this information was modulated in a phase-dependent manner and was more or less poorly related to the level and timing of stretch of the muscle-tendon units of the legs. Our interpretation of these experiments is that it is the "ensemble feature" of sensory information processing (Pearson, 1995; Prochazka and Gorassini, 1998a; 1998b) that provides important modulatory signals to the spinal cord. For example, the spinal cord seems to be processing all of the sensory information associated with the dynamics (kinetics and kinematics) of both lower limbs, the hips and the torso, perhaps from muscle, tendon, joint and/or skin receptors. We propose that this "load-sensing" reflects the interpretation of an ensemble of afferent information and cannot be attributed to any one type of sensory receptor, e.g., muscles spindle or Golgi tendon organ. Thus, the "perception" of load by the spinal networks reflects the combined effects of all receptors in some given state dependence. In other words, we propose that there are combinations of load receptors, not a single type of receptor, that generates a sense of loading.

The results from step-training studies (Barbeau, Pepin, Norman, Ladouceur and Leroux, 1998; Harkema, Hurley, Patel and Requejo, 1997; Wernig, Muller, Nanassy and Cagol, 1995) provide evidence that the human spinal cord can generate motor synergies reflecting motor behaviors that can be categorized as primary (formulated during development) and secondary (formulated from experience and learning) repertoires as described by Edelman (1987). Furthermore, these experiments demonstrate that the human lumbosacral spinal cord can assimilate, process and interpret complex proprioceptive information so that predictable motor repertoires such as stepping, standing and limb flexion can be performed. The execution of these motor tasks in spinal cord injured subjects having no supraspinal input demonstrate that the human spinal cord can dynamically organize cellular populations that can generate the essential synergies needed for stepping. It also appears that secondary repertoires can occur in the human spinal cord, since the synaptic efficacy of those networks that generate stepping and standing can be markedly improved with repetitive movements in which the appropriate sensory ensemble is appropriately matched, i.e. context dependent learning. These data are consistent with the Neuronal Group Selection theory in that it appears that stepping and standing can be performed by multiple neural pathways that are functionally, but not morphologically, equivalent networks. In the formulation of the concept of Neural Darwinism developed by Edelman (1987), the adaptive features which underlie this concept are assumed to be attributable to the brain. No suggestion or implication is made in his description that this concept may be applicable to the spinal cord. It appears to us, however, that essentially every underlying concept of Neural Darwinism is evident in spinal networks of animals including humans, as it appears to be in supraspinal structures.

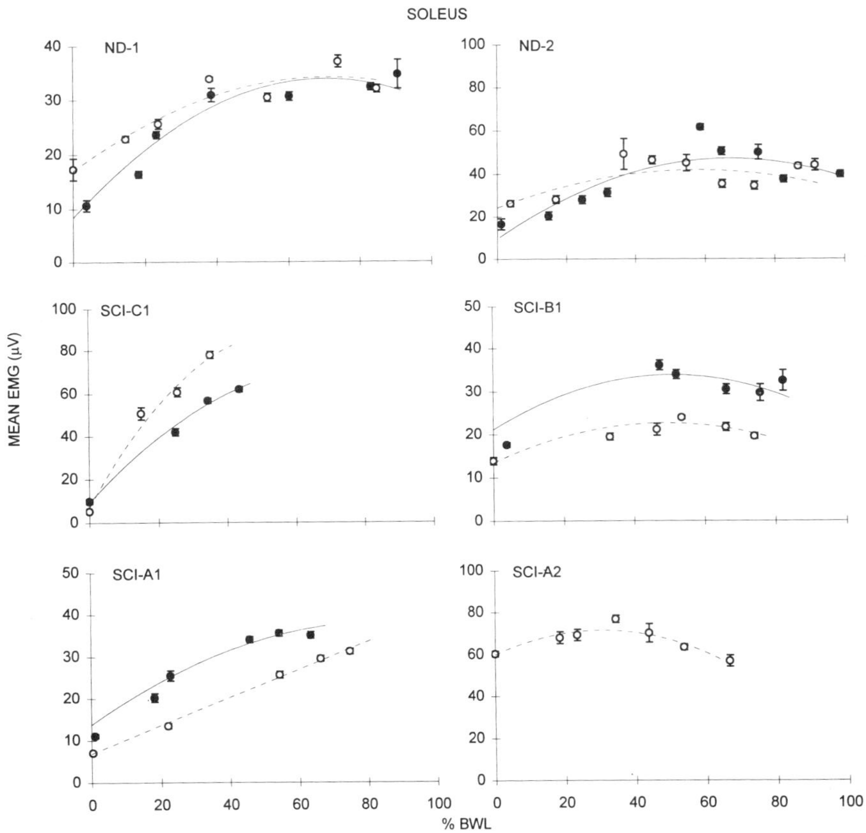

Figure 8.7. The subjects were suspended over a treadmill belt via a parachute-like vest. The level of loading relative to the subjects' body weight was determined from a load cell placed on the suspension cable and also via load sensors placed inside the shoes. Two non-disabled subjects (ND-1 and ND-2), two subjects categorized as ASIA B or C (SCI-B1 and SCI-C1), i.e., having minimal voluntary movement below the lesion, and two subjects categorized as ASIA A (SCI-A1 and SCI-A2), i.e. having no sensory or motor function crossing the spinal lesion. Other characteristics of the subjects are included in Harkema et al. (32). Averaged soleus EMG mean amplitude per burst (µV; rectified, high-pass filtered at 32 Hz) vs. limb peak load [% body weight load (BWL)] for each subject. Each point represents the average of the soleus EMG amplitudes (means ± SE) within a 10% interval of the BWL range (i.e., 1-10%, 10-20%, etc.). Data from the right limb (o) and the left limb (•) of each subject are shown separately. Data were not available for every BWL interval. The regression lines were based on individual data points from each step cycle and drawn for each muscle when there was a significant relationship (P<0.05).

8.10 ACKNOWLEDGEMENTS

This work was supported, in part, by NIH Grant NS16333, the Christopher Reeve Paralysis Foundation, the Kent Waldrep National Paralysis Foundation, NASA Grant NAG 2-717 and NAG2-1270. The authors would also like to thank Bruce Dobkin, Stephanie Elkins, Susan Harkema, Maynor Herrera, Robin Joynes, and Niranjala Tillakaratne for their assistance.

8.11 REFERENCES

Baev, Konstantin V. *Biological Neural Networks: The Hierarchical Concept of Brain Function,* Boston: Birkhauser, 1998.

Barbeau, H., Julien, C., and Rossignol, S. The effects of clonidine and yohimbine on locomotion and cutaneous reflexes in the adult chronic spinal cat. *Brain Research,* 1987; 437:83-96.

Barbeau, H., and Rossignol, S. Recovery of locomotion after chronic spinalization in the adult cat. *Brain Research,* 1987; 412:84-95.

Barbeau, H., and Rossignol, S. The effects of serotonergic drugs on the locomotor pattern and on cutaneous relfexes of the adults chronic spinal cat. *Brain Research,* 1990; 514(1):55-67.

Barbeau, H., and Rossignol, S. Initiation and modulation of the locomotor pattern in the adult chronic spinal cat by noradrenergic, serotonergic and dopaminergic drugs. *Brain Research,* 1991; 546:250-260.

Barbeau, H., Pepin, A., Norman, K., Ladouceur, M., and Leroux, A. Walking following spinal cord injury: control and recovery. *Neuroscientist,* 1998; 4:14-24.

Capaday, C, and Stein, R.B. Difference in the amplitude of the human soleus H relfex during walking and running. *Journal of Physiology (London),* 1987; 392:513-522.

Chau, C., Barbeau, H., and Rossignol, S. Effects of intrathecal 1- and 2- noradrenergic agonists and norepinephrine on locomotion in chronic spinal cats. *Journal of Neurophysiology,* 1998a; 79:2941-2963.

Chau, C., Barbeau, H., and Rossignol, S. Early locomotor training with clonidine in spinal cats. *Journal of Neurophysiology,* 1998b; 79:392-409.

de Guzman, C.P., Roy, R.R., Hodgson, J.A., and Edgerton, V.R. Coordination of motor pools controlling the ankle musculature in adult spinal cats during treadmill walking. *Brain Research,* 1991; 555:202-214.

de Leon, R.D., Hodgson, J.A., Roy, R.R., and Edgerton, V.R. Locomotor capacity attributable to step training versus spontaneous recovery following spinalization in cats. *Journal of Neurophysiology,* 1998a; 79:1329-1340.

de Leon, R.D., Hodgson, J.A., Roy, R.R., and Edgerton, V.R. Full weight-bearing hindlimb standing following stand training in the adult spinal cat. *Journal of Neurophysiology,* 1998b; 80:83-91.

de Leon, R.D., Hodgson, J.A., Roy, R.R., and Edgerton, V.R. The retention of hindlimb stepping ability in adult spinal cats after the cessation of step training. *Journal of Neurophysiology,* 1999a; 81:85-94.

de Leon, R.D., Tamaki, H., Hodgson, J.A., Roy, R.R., and Edgerton, V.R.. Hindlimb Locomotor and Postural Training Modulates Glycinergic Inhibition in the Spinal Cord of the Adult Spinal Cat. *Journal of Neurophysiology.* 1999b; 82:359-369.

Dietz, V., Colombo, G., Jensen, L., Baumgartner, L. Locomotor capacity of spinal cord in paraplegic patients. *Annals of Neurology,* 1995; 37:574-582.

Durkovic, R.D. The spinal cord: a simplified system for the study of neural mechanisms of mammalian learning and memory. In *Development and Plasticity of the Mammalian Spinal Cord,* ME Goldberger, A Gorio, M Murray (eds.) New York: Springer-Verlag,1986; 183-192.

Duysens, J.D., Clarac, F., and Cruse, H. Load-regulating mechanisms in gait and posture:comparative aspects. *Physiological Reviews,* 2000; 80(1): 83-133.

Edelman, G.M. Group selection and phasic reentrant signaling: A theory of higher brain function. In *The mindful brain: Cortical organization and the group-selective theory of higher brain function,* GM Edelman and VB Mountcastle (eds.) Cambridge, Mass.: MIT Press, 1978.

Edelman, G.M. Group selection as a basis for higher brain function. In *Organization of the cerebral cortex,* FO Schmitt, FG Worden, G Adelman, and SG Dennis (eds.) Cambridge, Mass.: MIT Press, 1981.

Edelman, G.M., Reeke, G.N., Selective networks capable of representative transformation, limited generalizations, and associative memory. *Proceedings of the National Academy of Sciences. USA,* 1982; 79:2091-2095.

Edelman, G.M., and Finkel, L.H. Neuronal group selection in the cerebral cortex. In *Dynamic aspects of neocortical function,* G.M. Edelman, W.E. Gall, and W.M. Cowan (eds.) New York: Wiley, 1984.

Edelman, G.M. *Neural Darwinism: The Theory of Neuronal Group Selection.* Basic Books, Inc., 1987.

Edgerton, V.R., de Guzman, C.P., Gregor, R.J., Roy, R.R., Hodgson, J.A., and Lovely, R.G. Trainability of the spinal cord to generate hindlimb stepping patterns in adult spinalized cats. In: *Neurophysiological Bases of Human Locomotion,* M. Shimamura, S. Grillner and V.R. Edgerton (eds.), Japan Scientific Societies Press, Tokyo, 1991, pp. 411-423.

Edgerton, V.R., Roy, R.R., Hodgson, J.A., Prober, R.J., de Guzman, C.P., and de Leon, R.D. Potential of adult mammalian lumbosacral spinal cord to execute and acquire improved locomotion in the absence of supraspinal input. *Journal of Neurotrauma,* 1991; 8:S119-S128.

Edgerton, V.R., Roy, R.R., Hodgson, Prober, R.A., de Guzman, C.P., and de Leon, R.D. A physiological basis for the development of rehabilitative strategies for spinally injured patients. *Journal of the American Paraplegia Society,* 1991; 14:150-157.

Edgerton, V.R., and Roy, R.R. Nervous System and sensory adaptation: neural plasticity associated with chronic neuromuscular activity. In: *Physical Activity, Fitness, and Health: International Proceedings and Consensus Statement.* Second edition, Chapter 34. C Bouchard, RJ Shephard, and T Stephens (eds.), Human Kinetics Publishers, Champaign, Illinois, 511-520, 1994.

Edgerton, V.R., de Leon, R.D., Tillakaratne, N., Recktenwald, M.R., Hodgson. J.A., and Roy, R.R. Use-dependent plasticity in spinal stepping and standing. In *Advances in Neurology: Neuronal Regeneration, Reorganization and Repair,* Philadelphia, PA: Lippincott-Raven Publishers, 1997a; 233-247.

Edgerton, V.R., Roy, R.R., de Leon, R.D., Tillakaratne, N., and Hodgson, J.A. Does motor learning occur in the spinal cord? *Neuroscientist,* 1997b; 3:287-294.

Forssberg, H., Grillner, S., and Rossignol, S. Phasic gain control of reflexes from the dorsum of the paw during spinal locomotion. *Brain Research,* 1977; 132:121-129.

Grillner S. Locomotion in the spinal cat. In *Control of Posture and Locomotion,* R,B, Stein, K.G. Pearson, R.S. Smith and J.B. Redford (eds.) New York: Plenum Press, 515-535, 1973.

Grillner S. Control of locomotion in bipeds, tetrapods, and fish. In *Handbook of Physiology, The Nervous System, Motor Control,* VB Brooks (ed.) Bethesda, MD: Am Physiol. Soc., 1179-1236, 1981.

Harkema, S.J., Hurley, S.L., Patel, U.K., and Requejo, P.S. Human Lumbosacral Spinal Cord Interprets Loading During Stepping. *Journal of Neurophysiology,* 1997; 77:797-811.

Hodgson, J.A., Roy, R.R., de Leon, R.D., Dobkin, B., and Edgerton, V.R. Can the mammalian lumbar spinal cord learn a motor task? *Medicine and Science in Sports Exercise,* 1994; 26:1491-1497.

Lovely, R.G., Gregor, R.G., Roy, R.R., and Edgerton, V.R. Effects of training on the recovery of full-weight-bearing stepping in the adult spinal cat. *Experimental Neurology,* 1986; 92: 421-435.

Lovely, R.G., Gregor, R.J., Roy, R.R., and Edgerton, V.R. Weight-bearing hindlimb stepping in treadmill-exercised adult spinal cats. *Brain Research,* 1990; 514:206-218.

Macpherson, J.M., and Fung, J. Weight support and balance during perturbed stance in the chronic spinal cat. *Journal of Neurophysiology,* 1999; 82(6):3066-3081.

Nakada, K., Hodgson, J.A., de Leon, R., Roy, R.R., and Edgerton, V.R. Prolonged modification of the mechanics of the step cycle by single and repetitive mechanical stimuli in chronic spinal cats. *Society for Neuroscience Abstracts,* 1994; 20:1755.

Orlovsky, G.N., Deliagina, T.G., and Grillner, S. *Neuronal Control of Locomotion: From Mollusc to Man,* Oxford University Press, 1999.

Pearson, K.G. Proprioceptive regulation of locomotion. *Current Opinion in Neurobiology,* 1995; 5:786-791.

Pierotti, D.J., Roy, R.R., Bodine-Fowler, S.C., Hodgson, J.A., and Edgerton, V.R. Mechanical and morphological properties of chronically inactive cat tibialis anterior motor units. *Journal of Physiology (London),* 1991; 444:175-192.

Pratt, C.A., Fung, J., Macpherson, J.M. Stance control in the chronic spinal cat. *Journal of Neurophysiology,* 1994; 71:1981-1985.

Prochazka, A., and Gorassini, M. Models of ensemble firing of muscle spindle afferents recorded during normal locomotion in cats. *Journal of Physiology,* 1998a; 507(1):277-291.

Prochazka, A., and Gorassini, M. Ensemble firing of muscle afferents recorded during normal locomotion in cats. *Journal of Physiology,* 1998b; 507(1):293-304.

Robinson, G.A., and Goldberger, M.E. The development and recovery of motor function in spinal cats. II. Pharmacological enhancement of recovery. *Exp\erimental Brain Research,* 1986; 62:387-400.

Rossignol, S., and Gauthier, L. An analysis of mechanisms controlling the reversal of crossed spinal reflexes. *Brain Research,* 1980; 182:31-45.

Rossignol, S., Barbeau, H., and Julien, C. Locomotion of the adult chronic spinal cat and its modification by monoaminergic agonists and antagonists. In *Development and Plasticity of the Mammalian Spinal Cord*, M. Goldberger, A. Gorio, and M. Murray (eds.). Padova: Fidia Research Series III, Liviana Press, 323-345, 1986.

Rossignol, S., and Barbeau, H. Pharmacology of locomotion: an account of studies in spinal cats and spinal cord injured subjects. *Journal of the American Paraplegia Society,* 1993; 16:190-196.

Rossignol, S. "Neural Control of Stereotypic Limb Movements." In *Handbook of Physiology Section 12. Exercise: Regulation and Integration of Multiple Systems*, L.B. Rowell and J.T. Shepherd (ed.) New York, New York: American Physiological Society, 1996.

Roy, R.R., Baldwin, K.M., and Edgerton, V.R. "The plasticity of skeletal muscle: effects of neuromuscular activity." In: *Exercise and Sport Science Reviews*. Baltimore, Williams and Wilkins, 269-312, 1991.

Roy, R.R., Hodgson, J.A., Lauretz, S.D., Pierotti, D.J., Gayek, R.J., Edgerton, V.R. Chronic spinal cord-injured cats: surgical procedures and management. *Laboratory Animal Science,* 1992; 42:335-343.

Roy, R.R., Eldridge, L., Baldwin, K.M., and Edgerton, V.R. Neural influence on slow muscle properties: inactivity with and without cross-reinnervation. *Muscle & Nerve,* 1996; 19:707-714.

Roy, R.R., Talmadge, R.J., Hodgson, J.A., Zhong, H., Baldwin, K.M., and Edgerton, V.R. Training effects on soleus of cats spinal cord transected (T12-13) as adults. *Muscle & Nerve,* 1998; 21: 63-71.

Selverston, A. "General principles of rhythmic motor pattern generation derived from invertebrate CPGs." In *Progress in Brain Researchearch Volume 123: Peripheral and Spinal Mechanisms in the Neural Control of Movement*, M.D. Binder (Ed.) Amsterdam, The Netherlands: Elsevier Science B. V., 1999.

Stein, P.S.G. "Central pattern generators and interphyletic awareness." In *Progress in Brain Research Volume 123: Peripheral and Spinal Mechanisms in the Neural Control of Movement*, M.D. Binder (ed.) Amsterdam, The Netherlands: Elsevier Science B. V., 1999.

Stewart, J.E., Barbeau, H., and Gauthier, S. Modulation of locomotor patterns and spasticity with clonidine in spinal cord injured patients. *Journal of Canadian Science and Neurology,* 1991; 18:321-332.

Tower, S.S. Function and structure in the chronically isolate lumbo-sacral spinal cord of the dog. *Journal of Comparative Neurology,* 1937; 67:109-131.

Weimann, J.M., and Marder, E. Switching neurons are integral members of multiple oscillatory networks. *Current Biology,* 1994; 4:896-902.

Wernig, A., Muller, S., Nanassy, A., and Cagol, E. Laufband therapy based on 'rules of spinal locomotion' is effective in spinal cord injured persons. *European Journal of Neuroscience,* 1995; 7:823-829.

9 SPINAL CORD PLASTICITY ASSOCIATED WITH LOCOMOTOR COMPENSATION TO PERIPHERAL NERVE LESIONS IN THE CAT

Laurent Bouyer and Serge Rossignol
Centre de recherche en sciences neurologiques
Pavillon Paul-G.-Desmarais
2960 Chemin de la Tour
Université de Montréal
Montréal (Québec), Canada H3T 1J4

9.1 INTRODUCTION

Effective rehabilitation of locomotion after lesions of the central or peripheral nervous system ultimately depends on the ability of remaining circuits to compensate for the loss of control mechanisms so as to re-express optimally residual locomotor functions. Whereas numerous studies have been devoted to try and define the roles played by the spinal cord, supraspinal structures and peripheral afferents in the control of locomotion, as reviewed several times (Armstrong, 1986; Grillner, 1981; Rossignol, 1996; Shik and Orlovsky, 1976), only a few studies have actually concentrated on the plastic changes that occur at various levels of the central nervous system to re-express functional locomotion after lesions of these structures. Studies on the recovery of hindlimb locomotion after a complete or a partial spinal transection at the last thoracic segment in cats have been reviewed recently (Rossignol, 1996; Rossignol, et al., 1996; Rossignol, et al., 1999). These studies indeed show a remarkable ability of the CNS to optimize the locomotor function with remaining deficits specific to each type of lesions. This ability to re-express locomotion undoubtedly reflects plasticity of locomotor controlling mechanisms which, depending on the lesion, may occur at different levels of the CNS.

This chapter will address the specific issue of functional plasticity in the locomotor system after various lesions and more specifically after peripheral nerve lesions (muscle nerves or cutaneous nerves). We will first review briefly some pertinent literature on locomotor training in spinal cats indicating that indeed the spinal cord is capable up to a certain extent to modify its output with training and therefore that the spinal cord has some locomotor adaptive potential. We will also review some studies on the adaptation of locomotion after tendon or nerve transfers which change the mechanical actions of muscles activated in the various phases of locomotion. The consequence of large deafferentation caused by dorsal rhizotomy or specific denervations will be briefly discussed. Finally, we will describe our own experiments aimed at comparing the adaptive locomotor capacities of normal and/or spinal cats to specific peripheral denervation in the hope of determining where in the CNS adaptation takes place and the limits of adaptation in different preparations

Locomotor training

The beneficial effects of locomotor training after a complete spinal transection suggest that the spinal circuitry responsible for generating locomotion may be, up to a certain extent, influenced by repetition. Although spinal animals will step after a few weeks without any special training, regular treadmill training will improve significantly the quality of the walk and the ability to support weight of the hindquarters during walking (Barbeau and Rossignol, 1987; Barbeau and Rossignol, 1994; Bélanger, et al., 1996; Dykman and Shurrager, 1956; Edgerton, et al., 1983; Lovely, et al., 1985; Lovely, et al., 1986; Smith, et al., 1982). Recent work also suggests that early training after spinalization, made possible by the intrathecal (i.t.) or intraperitoneal injection of clonidine, an alpha-2 noradrenergic agonist, may accelerate the appearance of weight-supported hindlimb locomotion (Barbeau, et al., 1993; Chau, et al., 1998a). Indeed, it is usually almost impossible to train efficiently a spinal cat to walk on the treadmill within the first week after spinalization because the animal is paralysed or else makes very small disorganized steps, contacting the belt with the dorsum of the foot without any significant weight support of the hindquarters. However, with a daily i.t. injection of clonidine (Barbeau, et al., 1987; Barbeau and Rossignol, 1991; Chau, et al., 1998b; Rossignol, et al., 1986), locomotion with plantar foot contact and ample movements can be induced on the treadmill for a period of 4 to 6 hours thus allowing us to effectively train the animal during that early post-spinal period. We found that in 4 cats, good locomotion could be obtained with plantar foot placement and weight support of the hindquarter after a week of training and thereafter the performance of these cats remained good for several weeks after without clonidine.

The issues of whether such training exerts its beneficial effects through central or peripheral mechanisms as well as the motor specificity of the training has been taken up by Edgerton and his colleagues (de Leon, et al., 1998b; de Leon, et al., 1998a; de Leon, et al., 1999; Edgerton, et al., 1991; Edgerton, et al., 1997; Hodgson, et al., 1994). They conclude that training exerts its effects mainly through central mechanisms and that cats trained specifically to walk will walk better than stand whereas cats trained to stand still will stand better than walk. Again, this suggests that, somehow, spinal neural circuits involved in particular tasks may be influenced by specific training because they are to some degree plastic.

The question of spinal "learning" has been at the centre of many of the presentations in this symposium. Is spinal locomotion learnt? Obviously, as mentioned above, spinal locomotion can be expressed very early after spinalization with noradrenergic drugs, either on a treadmill or in "fictive" conditions i.e. after paralysis (Forssberg and Grillner, 1973; Grillner and Zangger, 1979). It is therefore a well-established fact that the spinal cord per se is capable of generating a locomotor pattern without having to "relearn" walking. Furthermore, kittens spinalized before having learnt to walk, can walk very well with their hindlimbs on a treadmill for several months after spinalization (Forssberg, et al., 1980a; Forssberg, et al., 1980b), suggesting that locomotion is an innate pattern which does not require as such to be "learnt" for its basic expression. However, it is clear that some form of learning is at play when the expression of this pattern is modulated by training. Accelerating the rate of locomotor recovery after spinalization or favouring standing over walking are both forms of modulation of locomotor expression that result from training. Similarly, young spinal rabbits can be trained with a mechanical device to either favor an alternate or an in-phase coupling of the hindlimbs (Viala, et al., 1986). Again, since both alternate and in-phase coupling modes exist in the rabbit, the patterns are not as such learnt but one or the other mode of coupling becomes preferred after training. It is doubtful that the spinal cord, per se, would be capable of learning motor patterns that are not to some degree already extant. However, it is interesting to investigate the possibility that the spinal cord can modify, by itself in isolation from descending inputs, the expression of the locomotor pattern when having to deal with changes in sensory inputs or even changes in the peripheral neuro-muscular apparatus. This adaptation of spinal locomotion, made possible by repetitive training, represents in our minds a form of spinal learning.

Tendon/nerve transfers

In an early attempt to address the problem of locomotor plasticity, physiologists at the turn of the century transposed muscles or muscle nerves and observed how laboratory animals (mainly frogs and cats) compensated long-term after surgery. The rationale was very straight forward: can the central nervous system learn to change the timing of activation of muscles in order to generate proper locomotion after the mechanical role of the given muscle had been reversed, i.e. changed from a flexor to an extensor or from an extensor to a flexor or from being an extensor on one side to being an extensor on the other side (reviewed in Sperry, 1945). These authors reported "whole limb" responses and claimed that compensation was readily obtained, thereby implying that the locomotor system was extremely plastic. But, as pointed out (Sperry, 1945), these experiments suffered from several weaknesses. For instance, non-transposed muscles normally acting across the same joint as the transposed muscle(s) were left intact or else the novel mechanical coupling of non-transposed muscles, brought about by the surgical procedure, were not fully appreciated (as discussed in Taylor, 1936).

In a very influential set of experiments performed in rats, Sperry (1940; 1941; 1942) later showed that if rats are forced to use only the transposed muscles (by cutting muscle nerves to all other muscles around the same joint), the compensation described by previous investigators was not seen. These results led Sperry to propose an alternative explanation: compensation was the result of the taking over of unaffected muscles or by muscles acting across more proximal joints in what he called "tricks" learned by the animal. According to Sperry, these tricks were similar to the strategy used after joint amputation.

The general view at that point was therefore that central programming was fixed, and could not be influenced by changes in sensory feedback. For locomotion this means that the networks of interneurons responsible for the generation of locomotor movements were actually very resistant to changes, and that the locomotor circuitry was not plastic. More recently, the problem of muscle crossing in the cat hindlimbs was reinvestigated with combined EMG and kinematic recordings during locomotion (Forssberg and Svartengren, 1983). They transposed the proximal portion of the lower tendon of the medial gastrocnemius muscle (MG) with the distal part of the cut tibialis anterior (TA) tendon. Even after 3 years, one cat still showed locomotor deficits and never changed the timing of activation of MG from extension to flexion. They also concluded that the locomotor system was "hardwired".

However, when forearm tendon transfers were repeated by Sperry in monkeys (1947), some compensation was seen after substantial training. Yumiya et al. (1979) transposed wrist muscles from the forepaw of cats and showed gradual locomotor compensation. Eccles et al. (1960) looked for changes in the simplest connections between afferents and motoneurones long-term after antagonist nerve crossing, the monosynaptic EPSPs. In kittens 1-25 days old, they crossed the proximal end of the Peroneus (Per) muscle nerve with the distal end of the Medial Gastocnemius (MG) muscle nerve and visa versa. Twenty-one to 38 weeks later, they made intracellular recordings in Per and MG motoneurones and compared the effects of homonymous and heteronymous nerve stimulation on the generation of PSPs in these cells to control cats. Their results showed the following. First, after cross-union, 60-80% of the group I afferent fibres were lost permanently due to chromatolytic degeneration. Secondly, some new connections were made between previously antagonist muscles, but the change was incomplete. Thirdly, the motoneurones kept their Ia monosynaptic connections with their muscle of origin and also their Ia disynaptic inhibition from the original antagonists, that were now agonists. This work clearly showed that muscle transposition would induce some degree of plastic changes in the spinal connectivity, even though the authors tended to support Sperry's view. Luff and Webb (Luff and Webb, 1985) crossed Soleus and EDL muscle nerves and showed that the phase of activity of these muscles during locomotion became variable after crossing, which they attributed to the combination of a lack of proprioceptive information (see also, Cope et al., 1994) and/or "some degree of central remodelling."

In conclusion, locomotor adaptation to muscle or nerve transpositions across a joint represents a very demanding task which seems to be overcome only by higher vertebrates and may be restricted to the forelimbs or to partial changes in the hindlimbs. In most cases, animals seem to develop "tricks", recruiting other muscles and/or joints to produce more or less the same biomechanical function, thus suggesting a rather limited locomotor plasticity, in the classical sense. Below we will present other experiments in which the deficit to overcome is not so overwhelming and which may be achieved by using "tricks". We will also discuss the view that using these "tricks" might eventually lead in part to modifications of the spinal circuitry itself, and hence be compatible with the notion of spinal learning.

Deafferentation/denervation

Deafferentations through rhizotomy (sensory deficits) or muscle nerve lesions (sensorimotor deficits) have been used to study locomotor compensation. Goldberger (1977) has shown that after a complete unilateral deafferentation of one hindlimb by cutting all dorsal roots providing inputs to the spinal cord from that limb, cats never recovered the ability to walk properly with that limb. However, if only one dorsal root was left intact (Goldberger, 1983; 1988), the animal could almost completely compensate for the locomotor deficit caused by the sensory loss.

Wetzel et al. (1973) have shown that when ankle extensors that normally subserve different roles during posture and locomotion are cut, the remaining muscles will be used to compensate. Similarly, successive removal of ankle extensors in the rabbit (Stewart, 1937) shows a remarkable take over of the remaining muscles after the section of each one as revealed by the gradual decrease of the yield seen after removing each muscle.

In conclusion then, the literature is somewhat contradictory on the issue of plasticity in the locomotor system. This confusion may have a lot to do with terminology. There are indications of some plasticity in the locomotor control mechanisms but these might not, in general, be powerful enough to overcome very important mechanical challenges such as seen after tendon transfers. The indications of plasticity are strong enough however to pursue the question in different types of preparations as detailed in the following sections in which we report the recovery observed after lesions of specific peripheral nerves.

Experimental approach and general methodology

Cats were first selected for their willingness to walk regularly on a treadmill. They were then implanted under general anaesthesia with electromyographic (EMG) electrodes in several hindlimb muscles. After recovery, the cats were trained to walk again on the treadmill at constant speed. Then nerve lesions were made, also under general anaesthesia, and the cats were trained to walk on the treadmill after the neurectomy. This training could last several weeks in some cases until no further change in the recovery could be observed. At this point, they were spinalized under general anaesthesia at T13 and again trained to walk on the treadmill as was done before in other experimental series (Barbeau et al., 1987; Barbeau and Rossignol 1987; Bélanger et al., 1996; Chau et al., 1998a). This protocol was approved by our University Ethics Committee.

Except for experiments with Lateral Gastrocnemius-Soleus nerve lesions, all cats were neurectomized before spinalization. In all cases, neurectomies were also performed after spinalization to investigate the capacity of the spinal cord alone to adapt to such lesions and compare with the adaptation seen before spinalization.

The experimental strategy used thus allowed us on the one hand to determine whether plastic changes that must occur when adapting to a peripheral nerve lesion are carried over after a spinal section and, on the other hand, permitted to study purely spinal mechanisms of locomotor adaptation after peripheral nerve lesions.

9.2 RESULTS

Tibialis Anterior and Extensor Digitorum Longus lesions

The first study of this kind (Carrier et al., 1997) involved sectioning major ankle flexors Tibialis Anterior (TA) and Extensor Digitorum Longus (EDL) on one side of the animal. A few days after such a neurectomy, the deficits at the ankle and kinematic compensation during locomotion could hardly be seen by visual inspection alone. A detailed kinematic analysis of course revealed a major reduction of ankle flexion albeit not complete (compare Fig. 9.1A and 9.1B). It should be remembered that other muscles, such as the Peroneus longus (not recorded), can also produce some degree of ankle flexion (Young et al., 1993). EMG analyses showed that there was some increase in amplitude of the knee flexor Semitendinosus (ST) and/or the hip flexor Sartorius (Sart). This was enough to increase the knee and/or hip angles by a few degrees so as to clear the foot during swing. The trajectories of the foot (Fig 9.2) during swing in three cats show that, despite a reduction in ankle flexion, the foot was lifted securely above the surface of the treadmill belt.

Two of the cats were then spinalized at T13 and trained for several weeks. Surprisingly, and contrary to the usual locomotor performance of spinal cats (Barbeau and Rossignol, 1987; Bélanger et al., 1996; Chau et al., 1998a), these cats displayed a quite erratic walking after spinalization characterized by hyperflexions of the denervated limb, absence of weight support on that side and landing on the dorsum of the foot during stance as well as very asymmetrical step cycles (Fig. 9.1C). The walking cycle of the intact contralateral limb was more or less normal since it received very abnormal inputs from the denervated limb. In one cat, the sequence of events was reversed so that the cat was first spinalized and then neurectomized. After spinalization only, this cat walked regularly and symmetrically as was reported in other studies on spinal cats (Bélanger et al., 1996). After the neurectomy of the ankle flexors, the cat continued to walk symmetrically albeit with an absence of ankle flexion leading to foot drag and an increase in proximal flexors. We have not observed, however, the hyperflexion and disorganized asymmetrical hindlimb locomotor movements that were seen when cats were first denervated and then spinalized.

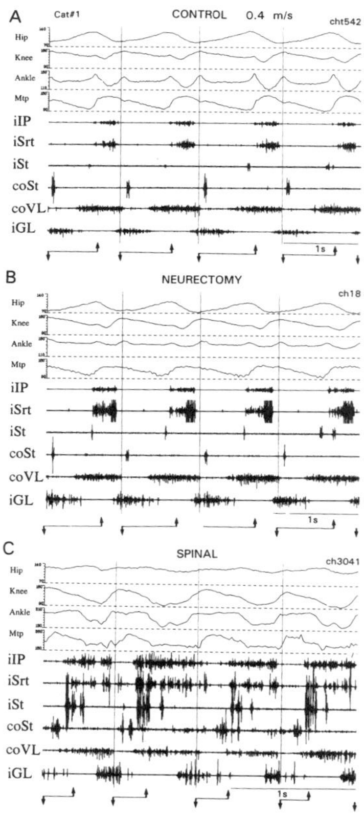

Figure 9.1. Kinematics and EMG activity of locomotion at 0.4m/s before and after neurectomy of TA and EDL in cat #1. The three panels have the same format. The 4 traces on top represent the angular movements. For the hip, knee and ankle, downward deflection of the traces indicate flexion whereas for the metatarsophalangeal jont (MTP), an upward deflection indicates plantar flexion in this figure only. Note that for the knee, the apparent flexion occurring during stance is due to a gradual slippage of the joint marker because of skin movements The EMGs are Ip: iliopsoas; Srt: Sartorius anterior; St: semitendinosus; VL, vastus lateralis; GL: gastrocnemius lateralis; i :ipsilateral to the lesion (left of the cat); co:contralateral to the lesion. In the duty cycles below the EMG traces, the downgoing arrows (9) indicate paw contacts, the upgoing arrows (8) indicate paw lifts and the horizontal lines represent the stance period. The vertical lines are aligned with the downgoing arrows. In A, 4 consecutive cycles from the control period are illustrated. In B, similar displays taken from 4 cycles 25 days post-neurectomy of TA and EDL. The gains of the EMG amplifiers were the same before and after neurectomy. In C, 4 consecutive cycles taken 19 days after spinalization of the neurectomized cat. Note the marked asymmetry between the two sides. Adapted with permission from (Carrier et al., 1997).

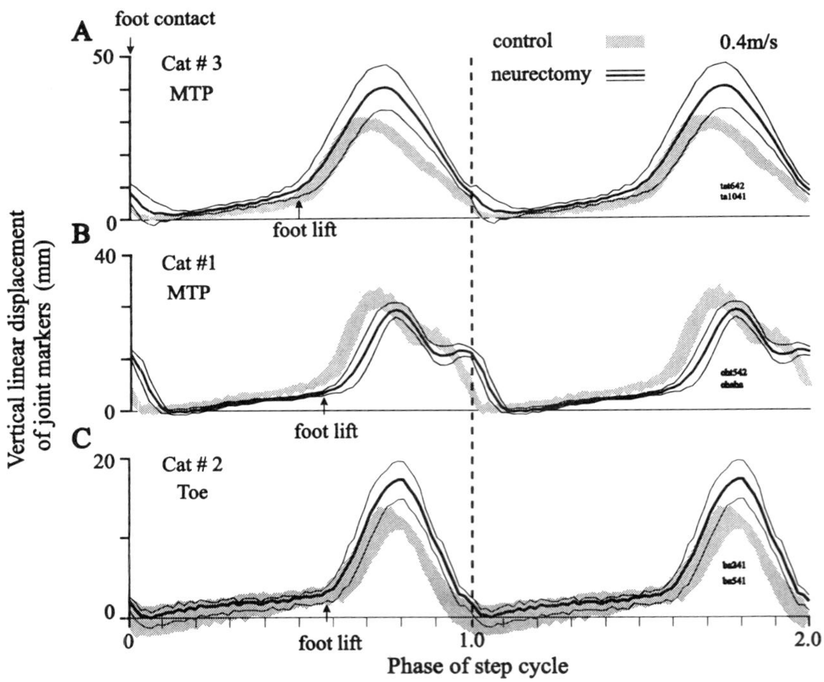

Figure 9.2. A,B,C Average vertical displacement of foot markers (toe in cat #1, metatarsophalangeal joint (MTP) in cats #2 and #3) during the normalized step cycle (about 10) at 0.4 m/s. The dotted line corresponds to foot contact whereas foot lift is indicated by arrows. In the 3 examples, The shaded curves display the trajectories of the foot (∀ 1 S.D.) in control conditions whereas the heavy line and thin lines represent the foot displacement (∀ 1 S.D.) after neurectomy. Note in cats #3 and 2 that there is a clear overshoot of the foot during swing. In cat #1 (see Fig.1), there is an increase flexion just before foot contact.

We concluded from these experiments that when a normal cat adapts to a denervation of the ankle flexors, changes occur at different levels of the CNS. Obviously there are changes at the spinal level since the expression of walking is quite different from that normally observed in spinal cats or when the neurectomy is performed after spinalization. Therefore some aspects of the locomotor spinal circuitry are changed and manifested as hyperflexions. It is possible that part of the compensatory strategy seen before spinalization (the increase knee and/or hip flexion) is achieved in part through an increase in the flexor component of the spinal locomotor generator. However this is not sufficient since the locomotor pattern is disorganized and the foot is not properly placed. We thus have to conclude that supraspinal control mechanisms play a most important part of the adaptive strategy seen before spinalization. This suggests that after such lesion, and maybe also normally, there is a step by step control over the spinally generated locomotor pattern.

Although we trained the one spinal cat neurectomized after spinalization for a few weeks, we could not see convincing evidence of improvements of locomotor adaptation in this situation. It is probable however that, had we severed the nerve only to TA, EDL could have compensated even in the spinal state as will be described in the next section with ankle extensors where agonists were left to permit some compensation.

Lateral Gastrocnemius-Soleus lesions

Pearson and collaborators (Whelan et al., 1995a; Whelan and Pearson, 1997) used another, complementary approach to study neurectomy-induced plasticity. They cut the muscle nerve innervating the two ankle extensors Lateral Gastrocnemius (LG) and Soleus (SOL), and measured the locomotor recovery of otherwise intact cats using kinematic and EMG recordings. As previously reported (Wetzel et al. 1973), they saw a rapid kinematic compensation. In addition, they showed that compensation involved an increased activity in at least one of the remaining ankle extensors MG, and most interestingly, this was accompanied by changes in polysynaptic reflex pathways (as tested indirectly by measuring effectiveness of increases in step cycle duration (Whelan et al., 1995b)). Furthermore, in 2/5 animals, these changes were retained after spinalization (Whelan and Pearson, 1997), suggesting that one of the compensatory mechanisms might be in the spinal cord. The next question then arises: are supraspinal structures necessary for this initial compensation to occur, or can a spinal animal also compensate for a similar deficit imposed after spinalization.

Together with Whelan and Pearson (Rossignol, et al., 1997), we have addressed this question. Spinal cats were trained to develop hindlimb locomotion and then were subjected to the same LGS nerve lesion as intact cats. Over the next few weeks, their hindlimb locomotion was then quantified using combined kinematic and EMG recordings. As expected, the nerve lesion caused a yield of the ankle during stance. Surprisingly, the kinematic locomotor recovery observed after the LGS lesion was similar, both in time course and in extent of recovery to that seen in intact cats (Rossignol, et al., 1997). Therefore, the spinal locomotor circuitry is capable of 1) detecting a muscle nerve lesion induced locomotor deficit and 2) compensating for it in an adequate manner in the absence of supraspinal assistance. Whether or not the same compensatory mechanisms are used by both preparations could not be addressed by this protocol.

Cutaneous denervation of the hindfeet

The two experiments presented above involved sectioning muscle nerves causing both sensory and motor deficits. In order to simplify further the study of spinal cord plasticity during locomotor compensation, we restricted our lesions to cutaneous nerves of the hindpaws (Bouyer and Rossignol, 1997; Bouyer and Rossignol, 1998; Bouyer and Rossignol, 1999). The unavoidable damage to the motor innervation of intrinsic foot muscles had only minimal functional consequences on the locomotor behaviour of the otherwise normal cat. The rationale and protocol were nearly identical to that described for the TA/EDL lesion, i.e. 1) compare the effects of the neurectomy performed before or after spinalization and 2) evaluate the residual capacities to compensate after a complete spinal transection.

Combined kinematics and EMG recordings were obtained from cats trained to walk regularly on a treadmill, before and after a complete, bilateral removal of cutaneous inputs from the hindlimb paws. The surgical procedure consisted of cutting at (or below) ankle level the following 5 cutaneous nerves: superficial peroneal (SP), tibial (Tib), caudal cutaneous sural (CCS), saphenous (Sap), and the cutaneous branch of the deep peroneal nerve (DPc). Only very small changes were visible in the kinematics of treadmill locomotion 2 days following this drastic change in sensory feedback in these otherwise intact cats. In Fig 9.3A, we see that the extension phase is slightly prolonged at the MTP joint (and sometimes at the more proximal joints) and the knee flexion slightly exaggerated. Averaged and rectified EMG recordings from representative muscles taken at the same time as the kinematics are shown in Fig. 9.3B. Changes were only seen in flexor muscles from the knee (Semitendinosus and Sartorius) and ankle flexors (not shown). No changes were seen in the EMG activity of extensor muscles. Even though these cats could walk almost normally on the treadmill, they could no longer walk across an horizontally placed ladder. Gradually, they however managed, within 25-51 days, to learn to walk across the ladder without any return of cutaneous feedback as tested by the absence of response to skin pinch. By that time, the kinematics of locomotion on the ladder had returned to normal but knee and ankle flexor EMG activity remained elevated.

If these changes occurred in the spinal cord itself, the compensation could be preserved after spinalization. Therefore denervated-compensated cats were spinalized and trained to walk on the treadmill, as we would have done for a spinal cat without such sensory deficits. Contrary to the non-denervated controls, spinal cats with cutaneous denervation of both hindpaws could not express a walking pattern with a plantar foot placement, early on (Fig. 9.4A) and even after more than a month after spinalization (Fig. 9.4B). Instead, they dragged their legs across the treadmill, keeping the dorsum of the foot in contact with the belt at all times (forward movement during swing and backward movement during stance) and performed rapid small amplitude movements. The pattern of muscle activation was normal except for ankle flexors which now discharged at the same time as ankle extensors, during the stance phase of locomotion (see TA and EDL in Fig. 9.4B). Apparently, the stretch signal from muscle afferents could not be used by the spinal cat to compensate for the missing cutaneous inputs.

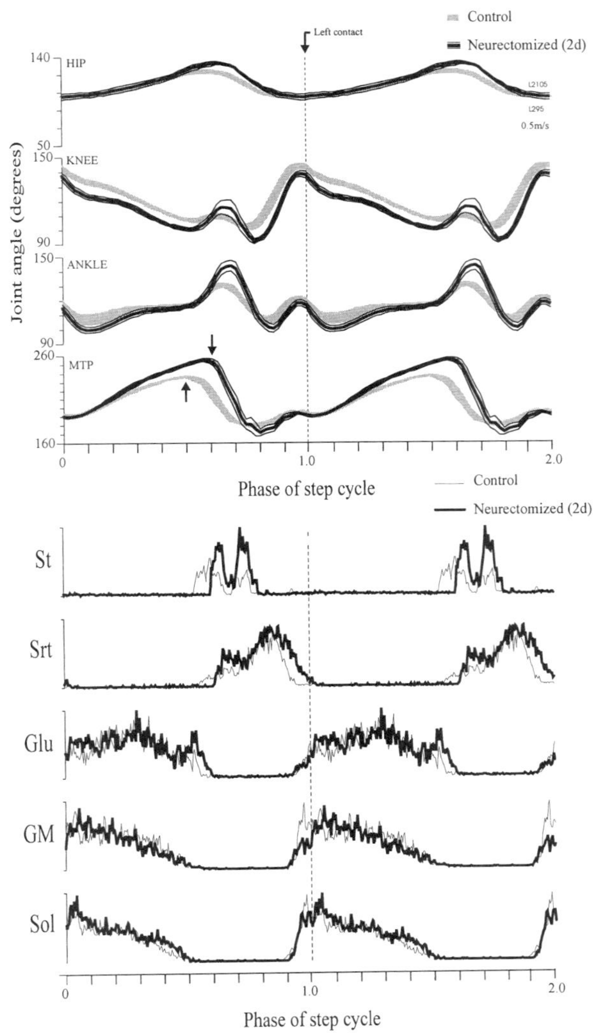

Figure 9.3. Top. Normalized angular displacements of the hip, knee, ankle and MTP joint in control (gray traces, n=10) and 2 days after neurectomy (black lines, n=10). Note that here, the flexion of MTP is represented by a downward deflection of the curve. Bottom. Averaged, rectified EMG from representative muscles obtained at the same time as the kinematics shown above. Glu: Gluteus; GM: gastrocnemius medialis; Sol: soleus. All data have been synchronized to foot contact (dashed line). Thin lines: control; thick lines: 2 days post-neurectomy.

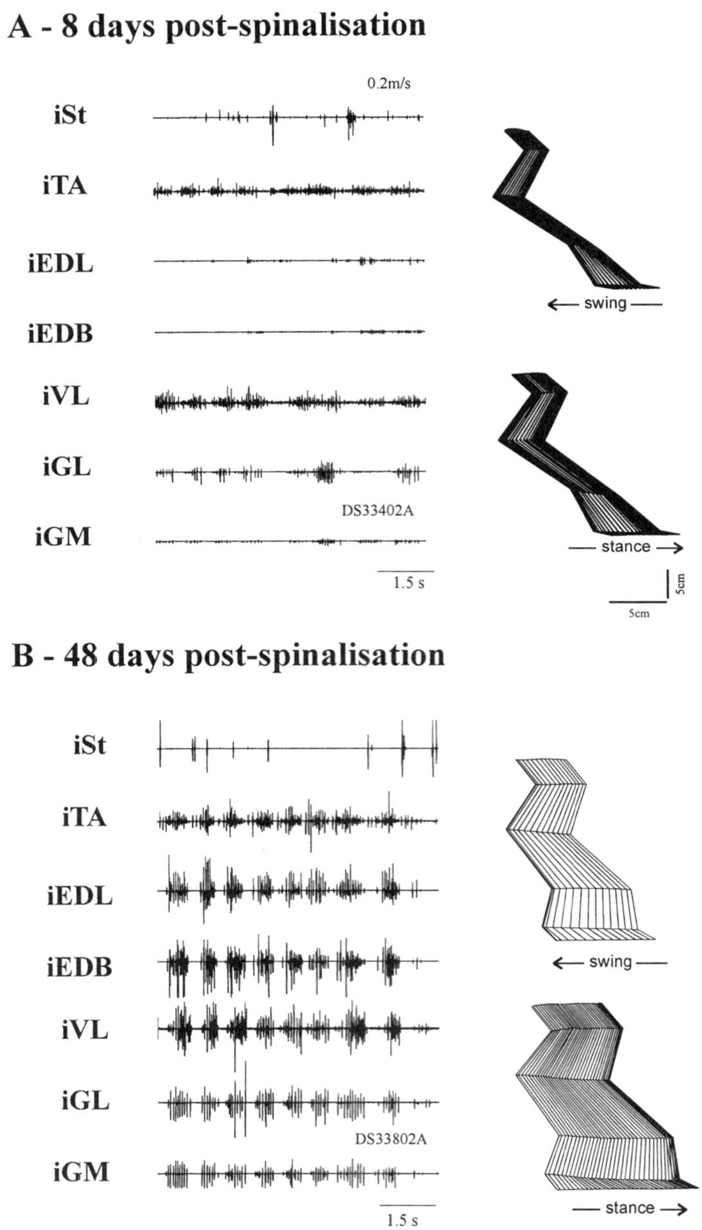

Figure 9.4. Raw EMG (left) and stick figure reconstruction of a step cycle (right) in a completely denervated cat. TA: tibialis anterior; EDL: extensor digitorum longus; EDB: extensor digitorum brevis; A. 8 days after spinalization. B. 48 days after spinalization. EMG activity in ankle (iTA, iEDL) and toe (iEDB) flexors is synchronous with that of ankle extensors (iGL, iGM) in these cats, contrary to their reciprocal activation in non-denervated controls (not shown).

In another chronic cat, we left one cutaneous nerve with a small receptive field intact, the DPc. Otherwise, the animal was put through the same experimental protocol as the completely denervated ones. Before spinalization, no difference was evident between this cat and the 2 others. After spinalization however, the cat learned to walk with plantar foot placement during assisted treadmill locomotion, similarly to (but more slowly than) the non-denervated controls (Fig. 9.5 A and B). This indicates a remarkable ability of the cat, even in the spinal state, to adapt its locomotion to a new constraint imposed by the sensory deprivation. This is what we consider as learning in the spinal cat; that is, the ability to modify, by repetitive training, the expression of the locomotor pattern so that it is best adapted to a change in the status of the peripheral apparatus (in this case a major reduction of sensory signals from the hindfeet). However this learning was not as exhaustive as in the denervated non-spinal case. After cutting the remaining cutaneous nerve, the cat permanently lost the ability to place the foot on the plantar surface, similarly to the cats which had had the complete cutaneous denervation in a single surgical procedure (Fig. 9.5 C).

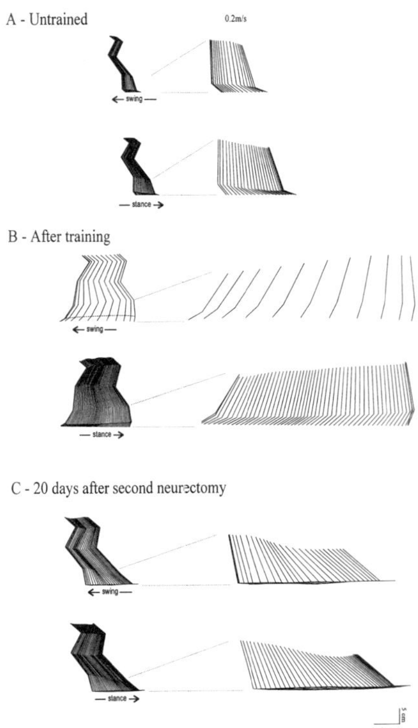

Figure 9.5. Partially denervated spinal cat. Left. Stick figure diagrams representing a complete step cycle. Right. Zooming on the ankle and foot to better show foot position. A. 3 days after spinalization and 79 days after partial neurectomy leaving DPc intact. B. 98 days after spinalization. C. 20 days after cutting DPc (161 days after spinalization).

9.3 CONCLUSIONS

In conclusion, locomotor compensation after lesions of motor (flexors or extensors) or cutaneous nerves occurs both in normal and spinal animals. In the normal cats such compensation occurs rapidly, within a few hours or days. This undoubtedly results from compensations by supraspinal mechanisms. How this is achieved (change in descending commands to motoneurones/interneurones, changes in gain of reflex pathways) is undetermined at this stage. Evidence provided by denervation of motor nerves suggests that there might be some long-term changes occurring the spinal cord itself. In the spinal state, cats can also adapt to partial ankle extensor denervation or partial cutaneous denervation. This adaptation occurs more slowly and presumably use different mechanisms than the intact cat. The spinal cat does not seem capable to adapt, even on a long-term basis, to a complete cutaneous denervation of the hindfeet. This indicates both that there are limits to what a spinal cat can adapt to. It also demonstrates that cutaneous inputs are essential to the expression of spinal locomotion. Studies are under way to investigate in these various models the contribution of various supraspinal structures and spinal mechanisms that might be involved in compensation in both intact and spinal cats.

9.4 REFERENCES

Armstrong, D.M. Supraspinal contributions to the initiation and control of locomotion in the cat. *Progress In Neurobiology,* 1986; 26: 273-361.

Barbeau, H., Chau, C., and Rossignol, S. Noradrenergic agonists and locomotor training affect locomotor recovery after cord transection in adult cats. *Brain Research Bulletin,* 1993; 30: 387-393.

Barbeau, H., Julien, C., and Rossignol, S. The effects of clonidine and yohimbine on locomotion and cutaneous reflexes in the adult chronic spinal cat. *Brain Research,* 1987;437: 83-96.

Barbeau, H. and Rossignol, S. Recovery of locomotion after chronic spinalization in the adult cat. *Brain Research,* 1987; 412: 84-95.

Barbeau, H. and Rossignol, S. Initiation and modulation of the locomotor pattern in the adult chronic spinal cat by noradrenergic, serotonergic and dopaminergic drugs. *Brain Research,* 1991; 546: 250-260.

Barbeau, H. and Rossignol, S. Enhancement of locomotor recovery following spinal cord injury. *Current Opinion in Neurology,* 1994; 7: 517-524.

Bélanger, M., Drew, T., Provencher, J., and Rossignol, S. A comparison of treadmill locomotion in adult cats before and after spinal transection. *Journal of Neurophysiology,* 1996; 76: 471-491.

Bouyer, L. and Rossignol, S. Hindfoot cutaneous input contribution to locomotion in intact and spinal cats. *Society for Neuroscience Abstracts,* 1997; 23(298.2), 759.
Ref Type: Abstract

Bouyer, L. and Rossignol, S. (1998). The contribution of cutaneous inputs to locomotion in the intact and the spinal cat. In Kiehn, O., Harris-Warrick, R. M., Jordan, L. M., Hultborn, H., and Kudo, N. (Eds.), *Neuronal Mechanisms for Generating Locomotor Activity* (pp. 508-512).

Bouyer, L. and Rossignol, S. Locomotor plasticity in the spinal cat after a progressive cutaneous denervation. *Society for Neuroscience Abstracts.* 1999.

Carrier, L., Brustein, L., and Rossignol, S. Locomotion of the hindlimbs after neurectomy of ankle flexors in intact and spinal cats: model for the study of locomotor plasticity. *Journal of Neurophysiology,* 1997; 77: 1979-1993.

Chau, C., Barbeau, H., and Rossignol, S. Early locomotor training with clonidine in spinal cats. *Journal of Neurophysiology,* 1998a; 59: 392-409.

Chau, C., Barbeau, H., and Rossignol, S. Effects of intrathecal α_1- and α_2-noradrenergic agonists and norepinephrine on locomotion in chronic spinal cats. *Journal of Neurophysiology,* 1998b; 79: 2941-2963.

Cope, T. C., Bonasera, S. J., and Nichols, T. R. Reinnervated muscles fail to produce stretch reflexes. *Journal of Neurophysiology,* 1994; 71: 817-820.

de Leon, R. D., Hodgson, J. A., Roy, R. R., and Edgerton, V. R. Locomotor capacity attributable to step training versus spontaneous recovery after spinalization in adult cats. *Journal of Neurophysiology,* 1998a; 79: 1329-1340.

de Leon, R. D., Hodgson, J. A., Roy, R. R., and Edgerton, V. R. Full weight-bearing hindlimb standing following stand training in the adult spinal cat. *Journal of Neurophysiology,* 1998b; 80: 83-91.

de Leon, R. D., Hodgson, J. A., Roy, R. R., and Edgerton, V. R. Retention of hindlimb stepping ability in adult spinal cats after the cessation of step training. *Journal of Neurophysiology,* 1999; 81: 85-94.

Dykman, R. A. and Shurrager, P. S. Successive and maintained conditioning in spinal carnivore. *Journal of Comparative and Physiological Psychology,* 1956; 49: 27-35.

Eccles, J. C., Eccles, R. M., and Magni, F. Monosynaptic excitatory action on motoneurones regenerated to antagonistic muscles. *Journal of Physiology,* 1960; 154: 68-88.

Edgerton, V. R., de Guzman, C. P., Gregor, R. J., Roy, R. R., Hodgson, J. A., and Lovely, R. G. (1991). Trainability of the spinal cord to generate hindlimb stepping patterns in adult spinalized cats. In Shimamura, M., Grillner, S., and Edgerton, V. R. (Eds.), *Neurobiological Basis of Human Locomotion* (pp. 411-423). Tokyo, Japan: Scientific Societies Press.

Edgerton, V. R., Johnson, D. J., Smith, L. A., Murphy, K., Eldred, A., and Smith, J. L. (1983). Effects of treadmill exercises on hindlimb muscles of the spinal cat. In Kao, C. C., Bunge, R. P., and Reier, P. J. (Eds.), *Spinal Cord Reconstruction* (pp. 435-444). New York: Raven Press.

Edgerton, V. R., Roy, R. R., de Leon, R. D., Tillakaratne, N., and Hodgson, J. A. Does motor learning occur in the spinal cord? *The Neuroscientist,* 1997; 3: 287-294.

Forssberg, H. and Grillner, S. The locomotion of the acute spinal cat injected with clonidine i.v. *Brain Research,* 1973; 50: 184-186.

Forssberg, H., Grillner, S., and Halbertsma, J. The locomotion of the low spinal cat. I. Coordination within a hindlimb. *Acta Physiologica Scandinavica,* 1980a; 108: 269-281.

Forssberg, H., Grillner, S., Halbertsma, J., and Rossignol, S. The locomotion of the low spinal cat: II. Interlimb coordination. *Acta Physiologica Scandinavica,* 1980b; 108: 283-295.

Forssberg, H. and Svartengren, G. Hardwired locomotor network in cat revealed by a retained motor pattern to gastrocnemius after muscle transposition. *Neuroscience Letters,* 1983; 41: 283-288.

Goldberger, M. E. Locomotor recovery after unilateral hindlimb deafferentation in cats. *Brain Research,* 1977; 123: 59-74.

Goldberger, M. E. (1983). Recovery of accurate limb movements after deafferentation in cats. In Kao, C. C., Bunge, R. P., and Reier, P. J. (Eds.), *Spinal cord reconstruction* (pp. 455-463) New York: Raven Press.

Goldberger, M. E. Spared-root deafferentation of cat's hindlimb: hierarchical regulation of pathways mediating recovery of motor behavior. *Experimental Brain Research,* 1988; 73: 329-342.

Grillner, S. Control of locomotion in bipeds, tetrapods, and fish. In Brookhart, J. M. and Mountcastle, V. B. eds. Handbook of physiology. The nervous system II. Bethesda, *American Physiological Society,* 1981; 1179-1236.

Grillner, S. and Zangger, P. On the central generation of locomotion in the low spinal cat. *Experimental Brain Research,* 1979; 34: 241-261.

Hodgson, J. A., Roy, R. R., de Leon, R. D., Dobkin, B., and Edgerton, V. R. Can the mammalian lumbar spinal cord learn a motor task? *Medicine and Science in Sports and Exercise,* 1994; 26: 1491-1497.

Lovely, R. G., Gregor, R. J., Roy, R. R., and Edgerton, V. R. Training effects on the recovery of full-weight-bearing stepping in adult spinal cats. *Society for Neuroscience Abstracts,* 1985; 11, 1101-1101.

Lovely, R. G., Gregor, R. J., Roy, R. R., and Edgerton, V. R. Effects of training on the recovery of full-weight-bearing stepping in the adult spinal cat. *Experimental Neurology,* 1986; 92: 421-435.

Luff, A. R. and Webb, S. N. Electromyographic activity in the cross-reinnervated soleus muscle of unrestrained cats. *Journal of Physiology (London),* 1985; 365: 13-28.

Rossignol, S. (1996). Neural control of stereotypic limb movements. In Rowell, L. B. and Sheperd, J. T. (Eds.), *Handbook of Physiology, Section 12. Exercise: Regulation and Integration of Multiple Systems* (pp. 173-216). Oxford: American Physiological Society.

Rossignol, S., Barbeau, H., and Julien, C. (1986). Locomotion of the adult chronic spinal cat and its modification by monoaminergic agonists and antagonists. In Goldberger, M. E., Gorio, A., and Murray, M. (Eds.), *Development and plasticity of the mammalian spinal cord* (pp. 323-345). Padova, Fidia Research Series III: Liviana Press.

Rossignol, S., Bouyer, L., Whelan, P. J., and Pearson, K. G. Chronic spinal cats can recover locomotor function following transection of an extensor nerve. *Society for Neuroscience Abstracts,* 1997; 23(298.14), 761.

Rossignol, S., Chau, C., Brustein, E., Bélanger, M., Barbeau, H., and Drew, T. Locomotor capacities after complete and partial lesions of the spinal cord. *Acta Neurobiologiae Experimentalis,* 1996; 56: 449-463.

Rossignol, S., Drew, T., Brustein, E., and Jiang, W. Locomotor performance and adaptation after partial or complete spinal cord lesions in the cat. Peripheral and spinal mechanisms in the neural control of movement. *Progress in Brain Research*, 1999.

Shik, M. L. and Orlovsky, G. N. Neurophysiology of locomotor automatism. *Physiological Reviews,* 1976; 56: 465-500.

Smith, J. L., Smith, L. A., Zernicke, R. F., and Hoy, M. Locomotion in exercised and non-exercised cats cordotomized at two or twelve weeks of age. *Experimental Neurology,* 1982; 76: 393-413.

Sperry, R. W. The functional results of muscle transposition in the hind limb of the rat. *Journal of Comparative Neurology,* 1940; 73: 379-404.

Sperry, R. W. The effect of crossing nerves to antagonistic muscles in the hind limb of the rat. *Journal of Comparative Neurology,* 1941; 75: 1-19.

Sperry, R. W. Transplantation of motor nerves and muscles in the forelimb of the rat. *Journal of Comparative Neurology,* 1942; 76: 283-321.

Sperry, R. W. The problem of central nervous reorganization after nerve regeneration and muscle transposition. *The Quarterly Review of Biology,* 1945; 20: 311-369.

Sperry, R. W. Effect of crossing nerves to antagonistic limb muscles in the monkey. *Archives of Neurology and Psychiatry,* 1947; 58: 452-473.

Stewart, D. Variations from normal gait after muscle section in rabbits. *Journal of Anatomy,* 1937; 72: 101-108.

Taylor, F. V. The effect of transposition of the achilles tendon on the walking and righting movements of the frog. *Comparative Psychology,* 1936; 21: 245-273.

Viala, D., Viala, G., and Fayein, N. Plasticity of locomotor organization in infant rabbits spinalized shortly after birth. In Goldberger, M. E., Gorio, A., and Murray, M. (Eds.), *Development and plasticity of the mammalian spinal cord* (pp. 301-310). Padova: Liviana Press, 1986

Wetzel, M. C., Gerlach, R. L., Stern, L. Z., and Hannapel, L. K. Behavior and histochemistry of functionally isolated cat ankle extensors. *Experimental Neurology,* 1973; 39: 223-233.

Whelan, P. J., Hiebert, G. W., and Pearson, K. G. Plasticity of the extensor group I pathway controlling the stance to swing transition in the cat. *Journal of Neurophysiology,* 1995a; 74: 2782-2787.

Whelan, P. J., Hiebert, G. W., and Pearson, K. G. Stimulation of the group I extensor afferents prolongs the stance phase in walking cats. *Experimental Brain Research,* 1995b; 103: 20-30.

Whelan, P. J. and Pearson, K. G. Plasticity in reflex pathways controlling stepping in the cat. *Journal of Neurophysiology,* 1997; 78: 1643-1650.

Young, R. P., Scott, S. H., and Loeb, G. E. The distal hindlimb musculature of the cat: multiaxis moment arms at the ankle joint. *Experimental Brain Research;* 1993; 96: 141-151.

Yumiya, H., Larsen, K. D., and Asanuma, H. Motor readjustment and input-output relationship of motor cortex following cross-connection of forearm muscles in cats. *Brain Research,* 1979; 177: 566-570.

10 LAUFBAND (TREADMILL) THERAPY IN INCOMPLETE PARA- AND TETRAPLEGIA

Anton Wernig, Andras Nanassy, and Sabina Müller
Department of Physiology
University of Bonn
Klinikum Karlsbad
Langensteinbach, Germany

10.1 SUMMARY

Recent reports indicate that intensive training of upright walking on a treadmill (German: Laufband, LB) according to the "rules of spinal locomotion", significantly improves walking capability in spinal cord damaged persons. The aids provided initially are body weight support by a harness and proper placing of one or both limbs by therapists.

The effects of this novel locomotion therapy on chronic and acute **incompletely paralyzed** patients, training methods and patient selection are summarized and discussed here. Many chronic patients still wheelchair-bound and not capable of walking without help by other persons, became independent and learned to walk for some distance without help. Training induced increase in voluntary muscle activity, its facilitation by proper limb and body posture, and involvement of complex locomotor reflexes (motor programs) are main sources for the improvements in locomotion.

Follow-up assessments performed 0.5 to 6.5 years after discharge from the hospital show that the significant improvements achieved by LB-therapy in initially chronic patients can be maintained under domestic surrounding. Initially acute patients further improved their walking capabilities.

It is suggested, that LB-therapy is generally applied in motor rehabilitation of acute and chronic incomplete para- and tetraplegic persons. Its use in other neurological diseases is discussed.

10.2 INTRODUCTION

In non-human primates, relearning of treadmill-stepping was described after most severe spinal cord damage, e.g. sparing only parts of the ventral white matter (Eidelberg, et al. 1981). In lower vertebrates like cat, such motor learning may occur even after complete transection of the low thoracic spinal cord (Lovely, et al. 1986; Barbeau, Rossignol, 1987; Edgerton, et al. 1991), indicating the existence of spinal locomotor programs capable, together with afferent information, of generating steplike movements in hindlimbs. In man after complete spinal transection, elements of stepping but no full step cycles may be evokable on the treadmill (Wernig, Müller, 1992; Wernig, et al. 1995; Wernig, Müller, 1996). Even with little voluntary activity remaining, however, full stepping could be entrained (Wernig, Müller, 1992; Wernig, et al. 1995). Improvement of stepping in spinal cord injured cats and monkeys was achieved by intensive treadmill-stepping; other training regimens like daily standing were not successful and even significantly worsened stepping capabilities in cats (Edgerton, et al. 1991). Translated into strategies for motor rehabilitation of spinal cord injured patients, these observations imply that relearning of walking can be done only by intensive exercise of upright walking.

Two independent groups of researchers have recently adopted aided treadmill locomotion (which in connection with humans we have called Laufband therapy), for spinal cord injured persons (Visintin, Barbeau, 1989; Wernig, Müller, 1991; Barbeau, Blunt, 1991; Wernig, Müller, 1992; Wernig, et al. 1995) and have reported improvements in locomotion. In a recent study the applicability of Laufband (LB) therapy, which focuses more on the "rules of spinal locomotion" (see below) than on body weight support, was investigated in 89 incompletely paralyzed chronic and acute para- and tetraplegics. The comparison with 64 patients, treated conventionally and both therapies applied consecutively in individual patients, showed a considerable efficacy of LB therapy for restoration or improvement of walking over ground (Wernig, et al. 1995). Improvements in locomotion achieved by LB therapy had occurred under controlled clinical conditions, usually within a few weeks only. Furthermore, as will be shown in detail, several patients had significantly improved their walking capabilities with little increase in voluntary muscle activity (Wernig, et al. 1995). Therefore, it was particularly important to know, and will be shown here, whether patients could maintain their new walking capabilities in domestic surroundings without daily LB therapy (Wernig, et al. 1998).

10.3 METHODS AND PATIENTS

The principles of LB therapy have been published before (Wernig, Müller, 1991; Wernig, Müller, 1992; Wernig, et al. 1995). In short, exercise of upright walking is performed on a motor driven treadmill. Patients are supported by a harness (Bonmed) suspended from the ceiling. The harness is critical as it needs to support the patient without blocking his/her ability to move, e.g. by forcing him into hip flexion. The amount of body weight support (BWS) is adjusted individually for each patient. Chronic patients already walking often do not tolerate more than 10-20% BWS, while acute patients initially may need 40-50% BWS.

Initially passive movements of the limbs are performed by 1-2 therapists, if necessary. For this purpose the treadmill needs to be elevated for therapists sitting at the sides (Figure 10.1).

Laufband (Treadmill)

Figure 10.1. Schematic depiction of locomotion therapy on a treadmill.

During walking "rules of spinal locomotion" are enforced by therapists to facilitate locomotion whenever necessary and appropriate; these rules, derived from animal experiments (Grillner, 1981; Lovely, et al. 1986; Barbeau, Rossignol, 1987; Wernig, Müller 1992; Edgerton, et al. 1991) imply; extension of hip and knee joints with optimal limb loading during stance; unloading and shift of body weight onto the contralateral limb to initiate swing; afferent stimulation like skin irritation by pinching or pressing supports to initiate the movements (Wernig, Müller, 1992; Wernig, et al. 1995). The use of special treadmill shoes, composed of a thin leather sole and with grips easy for the therapist to handle, are of great benefit especially in the initial phase of therapy. LB therapy was performed usually once, in some patients 2 times per day, for periods of 30 minutes for 5 days a week. Walking training on firm ground should also be exercised as soon as possible. Apart from this, patients participated in the regular conventional rehabilitation program for indoor patients, however, walking initially was performed only under control to ensure use of the "rules of spinal locomotion".

Criteria for selecting patients to enter LB therapy are: Presence of some voluntary muscle activity in the lower limbs in particular of quadriceps femoris, mobility of joints, no severe muscle shortenings and no skin ulcerations or other severe diseases. Missing voluntary hip flexion can be tolerated initially especially if it can be elicited by facilitating measures in the initial testing.

Start of therapy: in the chronic patients reported here, LB therapy started 6 months to 15 years after spinal cord injury (median 1.5 years). In acute patients who had suffered trauma of the spinal column, the safety of the procedure has to be assured by the orthopaedic surgeon; due to surgical stabilization of the vertebral column (Harms, 1992), start of walking exercise and other physical therapy usually was allowed within a few weeks after trauma.

Assessments

Walking capabilities before and after LB therapy as well as in the follow-up investigations were classified into 6 classes (0-5) solely based on the patient's ability to walk on even ground (Table 10.1). Accordingly, wheelchair-bound patients not capable of standing up and walking without help are separated from those who are not wheelchair-bound.

Table 10.1 Classification of walking capability

Wheelchair-bound:

0 = lower limbs cannot support body weight for standing or walking, even with moderate help by 2 therapists

1 = standing and walking possible with the help of two therapists;

2 = walking at the railing with help of one therapist.

Not wheelchair-bound:

3 = walking with the rollator/walker or reciprocal frame without help;

4 = walking with two regular canes or four-point canes;

5 = walking without devices (free walking) for more than five steps.

For the initial rating a distance of 5-10 meters was evaluated, but 50-100 meters for the later ratings unless stated differently. These assessments were done by two independent evaluators (SM and AM) from the video film records; in case of discrepancies the higher class was assumed before start of therapy and the lower class later on.

The amount of voluntary muscle activity was evaluated from the force and range of movements evoked upon verbal command in defined resting positions (horizontal and sitting) avoiding readily evokable spastic extension movements (Kendall, et al. 1971).

Table 10.2 Classification of voluntary muscle activity

0 =	no muscle contraction visible or palpable
1 =	muscle contraction visible and/or palpable, no movement of limbs
2 =	some joint angle movement with passive support by the therapist balancing gravity
3 =	full range of joint angle movement against gravity
4 =	full movement plus maintenance of position against moderate applied resistance
5 =	like 4, against maximal applied resistance.

Values in between were allowed and valued as 0.5 points.

As an overall measure for muscle function in a limb, values of the following 8 major flexor and extensor muscles/muscle groups were summed up: glutaeus maximus, glutaeus medius and minimus, iliopsoas, sartorius, quadriceps femoris, ham strings, tibialis anterior, triceps surae. These values are depicted in Figure 10.4 for each leg separately.

Follow-up investigations were performed the same day patients arrived at the hospital for routine ambulatory check-ups or within the first days in the hospital in case of renewed indoor treatment. Patients were asked to walk over even ground at the optimal speed with their preferred devices used in their domestic surrounding. Videorecordings were made from all performances.

Assesssments of walking capability before and after treadmill therapy from 21 of the 35 chronic patients reported here in detail are included in a previous report (Wernig, et al. 1995) (patients coded 0/D-0/J, 1/A and 1/H and 1/H and 2/A-2/F in Table 10.4). This is also the case for 22 of the acute patients in Figure 10. 3.

The cause of spinal cord injury most frequently was trauma followed by non-progressive myelitis, tumours, vascular disorders and other causes (see Table 10.3). Chronic patients with components of flaccid paralyses are not included in this report.

10.4 RESULTS

Initially chronic patients

Walking capabilities of pre- and post-LB therapy of 35 para- and tetraplegic patients with spastic paresis are summarized in Figure 10.2 (upper and middle histograms). Out of 25 wheelchair-bound patients (classes 0-2, black bars in the upper histogram in Figure 10.2) 20 became independent after LB therapy and only 5 remained the same (middle of histogram). Patients already independent before LB therapy (classes 3-5, open bars in the upper histogram in Figure 10.2) ususally remained within their original functional class (only two patients advanced to higher classes), but all improved in speed and endurance of walking (data not shown).

Characterization of patients

Table 10.3: Different parameters of 25 chronic , intially wheelchair bound patients who underwent LB-therapy

Code	age	sex	cause of damage [1]	segment	start LB after s.c. damage	duration of LB therapy
0/A	62	f	T	Th 6	10 months	20 weeks
0/B	25	m	T	C 5	8,5 months	18 weeks
0/C	25	m	T	Th 7	6 years	12 weeks
0/D	67	m	T	C 6*	5,5 months	18 weeks
0/E	22	f	M	C 5*	12 months	16 weeks
0/F	50	f	T	C 5*	12 months	11 weeks
0/G	21	m	T	C 7*	22 months	10 weeks
0/H	52	m	Tu	Th12	20 months	9 weeks
0/I	49	m	T	C 5 **	6 months	20 weeks
0/J	30	m	T	C 5 **	6 months	20 weeks
1/A	19	f	T	Th10	11 months	12 weeks
1/B	56	f	V	Th 6	2 years	18 weeks
1/C	56	m	T	C 5*	4 years	11 weeks
1/D	35	m	T	Th 7	15 months	8 weeks
1/E	56	m	V	Th 3	12 months	14 weeks
1/F	51	m	T	C4*	4 years	13 weeks
1/G	24	m	T	C5	12 months	10 weeks
1/H	50	m	T	C6**	15 years	8 weeks
2/A	70	f	Tu	C 8	3 years	9 weeks
2/B	29	f	T	Th10	10 months	8 weeks
2/C	50	m	T	C 5	5,5 months	18 weeks
2/D	55	m	M	C6	11 months	10 weeks
2/E	28	f	T	C 7*	12 months	16 weeks
2/F	30	m	T	C4*	13 months	14 weeks
2/G	49	m	T	C 5*	12 months	10 weeks

1 M = myelitis, T = trauma, Tu = tumour, V = vascular disorder

Median duration of LB therapy was dependent on the severeness of the paralysis. It was 8-20 weeks (median 12 weeks) for wheelchair-bound patients and 4-12 weeks (median 10 weeks) for the others. Antispastic medication applied in 27 of the 35 patients was maintained in 8 but could be reduced in the course of LB therapy in 19.

Marked increases in voluntary muscle activity evoked under LB therapy occurred only in 5 patients (coded 0/A, 0/D left limb, 0/F, 2/B and 2/D in Table 10.4); noteworthy, 4 of these 5 patients were among those in which the antispastic treatment was markedly reduced, highlighting the well-known blocking effects of such drugs on voluntary muscle activity (full dots in Figure 10.4 upper part).

Strikingly on the other hand, in spite of improved locomotion, one or both limbs often remained nearly completely paralyzed, i.e. had cumulated values of 10 and lower for the 8 major flexor and extensor muscles in this limb (see Methods), with no single muscle exceeding 2.5 (patients coded 0/C, 0/D, 0/E, 0/G, 0/I, 1/A, 1/F, 1/G, 1/H and 2/G in Table 10.4).

The follow-up investigations performed 0.5 to 6.5 years later (median 20 months) show that the improvements achieved under clinical conditions by LB therapy were generally maintained (Figure 10.2, lower histogram and Table 10.4). Only one patient fell back from class 3 to 2 while one patient improved from 2 to 3 and 2 patients improved from 3 and 4 to 5. With the remarkable exception of 2 patients, the voluntary muscle activity was similar to that at discharge from the hospital (Figure 4 lower part). However, 2 patients (patients coded 0/I and 2/C in Table 4) showed a dramatic increase in cumulative values (from 10 and 7.5 up to 20.5 and 18.5; and from 27.5 and 16 up to 29.5 and 19.5). Antispastic medication was not varied during this period or only slighty, therefore it seems that the increase in voluntary muscle activity was due to improvement in the efficacy of descending tracts.

Initially acute patients

Forty-one para- and tetraplegic patients had started LB therapy 3-16 weeks (median 8 weeks) after spinal cord damage and performed it as part of their postacute rehabilitation for periods of 3-22 weeks (median 10 weeks). Figure 10.3 (upper and middle histograms) compares their walking capability at the beginning and end of LB therapy; according to this, 8 patients were still wheelchair-bound (classes 0-2) after the end of their first rehabilitation.

At the follow-up investigation 6 months to 6 years after discharge from the hospital (median 17 months), only six patients had remained wheelchair-bound. In addition, fewer patients needed a walker (class 3), clearly more patients could walk for short distances also without any devices (class 5) (lower histogram). In summary, of the initially acute patients 15 further improved and none of them reduced his/her walking capability after discharge from the hospital.

Table 4: Motor capabilities before and following Laufband therapy and at follow up investigations 25 chronic, intially non-independent walkers

Code	walking capabilities 2			stair case walking 3			cumulative voluntary muscle activity 4					
							right limb			left limb		
	before LB	after LB	follow up	before LB	after LB	follow up	before LB	after LB	follow up	before LB	after LB	follow up
0/A	0	4	5	no	yes	yes	13	26	29.5	19.5	30	32.5
0/B	0	3	3	no	yes	no	4.5	10.5	-	5	8.5	-
0/C	0	3	3	no	yes	yes	3	4	-	2.5	3.5	-
0/D	0	3	3	no	no	no	1	4	4.5	5	18.5	18.5
0/E	0	3	3	no	no	no	24	29.5	29.5	0	0	0
0/F	0	5	5	no	yes	yes	15	23.5	23.5	18.5	26.5	27.5
0/G	0	2	2	no	no	no	8.5	12.5	15	5	7	7.5
0/H	0	2	2	no	no	no	14	17	-	16	17.5	-
0/I	0	2	3	no	no	no	7	10	20.5	5.5	7.5	18.5
0/J	0	1	1	no	yes	yes	8	13.5	-	10	15	-
1/A	1	4	4	no	yes	yes	4	5.5	9	15	15.5	17.5
1/B	1	4	4	no	yes	yes	8.5	10.5	13	24	25.5	27
1/C	1	3	3	no	yes	yes	13.5	14	16.5	12.5	13	14.5
1/D	1	3	3	no	no	no	5.5	7	-	11.5	14.5	-
1/E	1	3	3	no	no	no	13.5	16.5	19.5	13	17.5	16.5
1/F	1	3	2	no	no	no	7.5	8	-	11.5	15	-
1/G	1	3	3	no	no	no	15	20.5	-	8.5	9	-
1/H	1	2	2	no	no	no	11	11	9.5	4.5	5.5	6.5
2/A	2	5	5	no	yes	yes	19	21.5	-	20	22.5	-
2/B	2	5	5	no	yes	yes	8	17.5	19	5	10	17
2/C	2	5	5	no	yes	yes	24	27.5	29.5	7.5	16	19.5
2/D	2	4	4	no	yes	-	12.5	23.5	-	13.5	24	-
2/E	2	4	4	no	yes	yes	27.5	31	28.5	7	14.5	13
2/F	2	4	4	no	yes	yes	6	10	13	29	30	30
2/G	2	3	3	no	yes	yes	6.5	6	7	8	12.5	15

2 Walking capability: 6 main classes defined in the text (s. table..)
3 Stair case walking with or without supporting persons.
4 Cumulative force of 8 major flexors and extensor determined in resting position (see Methods)

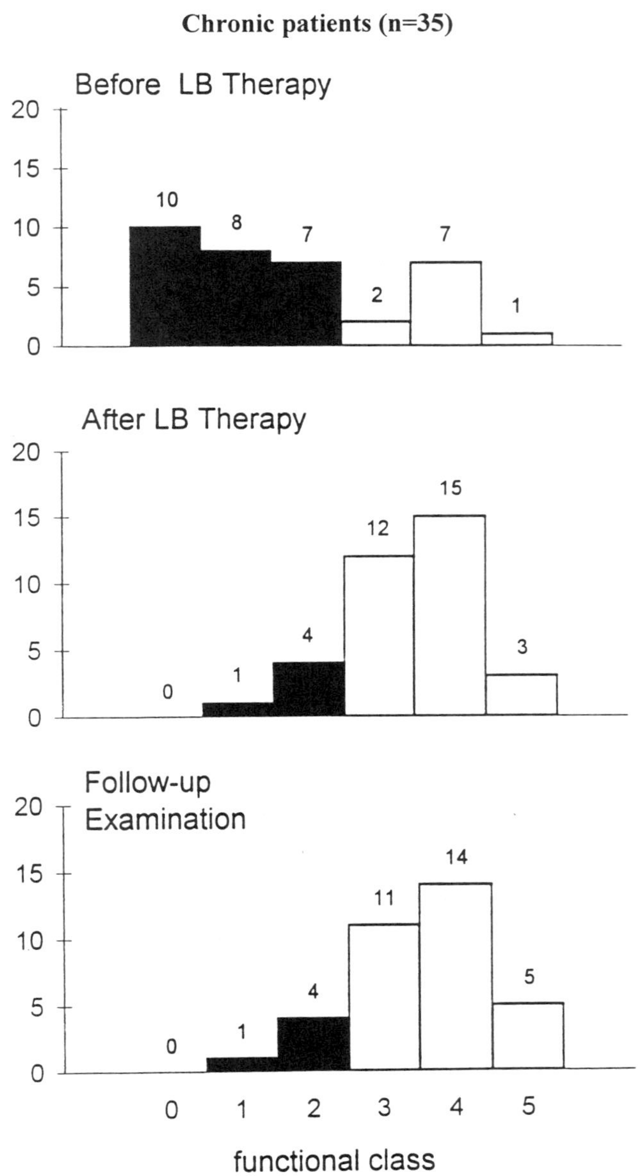

Figure 10.2. Walking capabilities according to 6 functional classes (0-5) in 35 chronic para- and tetraplegic patients before (upper histogram) and immediately after Laufband (LB) therapy performed in the hospital (middle histogram) as well as 0.5 to 6.5 years (median: 20 months) after discharge from the hospital (lower histogram, follow-up investigation). Black bars: wheelchair-bound patients. White bars: not wheelchair-bound patients (see Table 10.1). Numbers of the columns give numbers of patients in this class.

At the follow-up investigation 6 months to 6 years after discharge from the hospital (median 17 months), only six patients had remained wheelchair-bound. In addition, fewer patients needed a walker (class 3), clearly more patients could walk for short distances also without any devices (class 5) (lower histogram). In summary, of the initially acute patients 15 further improved and none of them reduced his/her walking capability after discharge from the hospital.

10.5 DISCUSSION

This report summarizes previous findings (Visintin, Barbeau, 1989; Barbeau, Blunt, 1991; Wernig, Müller, 1991, 1992, 1995; Wernig, et al. 1995) that spinal cord injured patients may improve their walking capability significantly by intensive training of upright walking on a treadmill. Improvement in the walking ability in some chronic patients went along with a surprisingly low increase of voluntary activity in resting position.

It seems, that the voluntary muscle activity during walking is facilitated by correct proprioceptive input, thus explaining the effectiveness of the "rules of spinal locomotion". This phenomenon was most striking in near-completely paralyzed patients. Patient coded 0/C, learned to walk over ground for distances of 20-40 meters with the help of a rollator, though his voluntary activity in the lower limbs did not exceed 4 and 3.5. Similar stepping was observed previously in a patient with zero voluntary activity in one limb (assessed also electromyographically) and a total value of 2 in the other (Wernig, Müller, 1992; Wernig, Müller, 1995). Obviously, training stabilizes the influence of proprioceptive information on the gait cycle. One of the features of the "rules of spinal locomotion", limb loading during stance, has recently been shown to dramatically enhance motor output from muscles of this limb in persons with complete (and incomplete) spinal cord transections (Maegele, et al. 1996; Harkema, et al. 1997). Therefore, it seems that even the completely isolated spinal cord correctly interpretes proprioceptive afferent information in the sense of maintaining upright body position. Applying these rules, short of complete step cycles, elements of stepping with hip flexion and knee extension, could recently be evoked in a 13-year old girl completely paralyzed below spinal segment Th 6 ((Wernig, Müller, 1996), see also Wernig, Müller, (1992) and Wernig, et al. (1995) for further reports on elements of stepping after complete spinal cord transection, and Dobkin, et al. (1995) and Dietz, et al. (1995) for phasic and alternating flexor and extensor muscle activity during aided stepping). Why then is a completely paralyzed patient not able to perform a complete stepping sequence like a cat does? Do we have to exercise more extensively? The 13-year old patient, described above, was trained on the treadmill two times a day over a period of 4 months, without becoming able to walk without help. It appears, that more of supraspinal inputs are needed in primates than in lower vertebrates to bring spinal motor programs above threshold.

Acute patients (n=41)

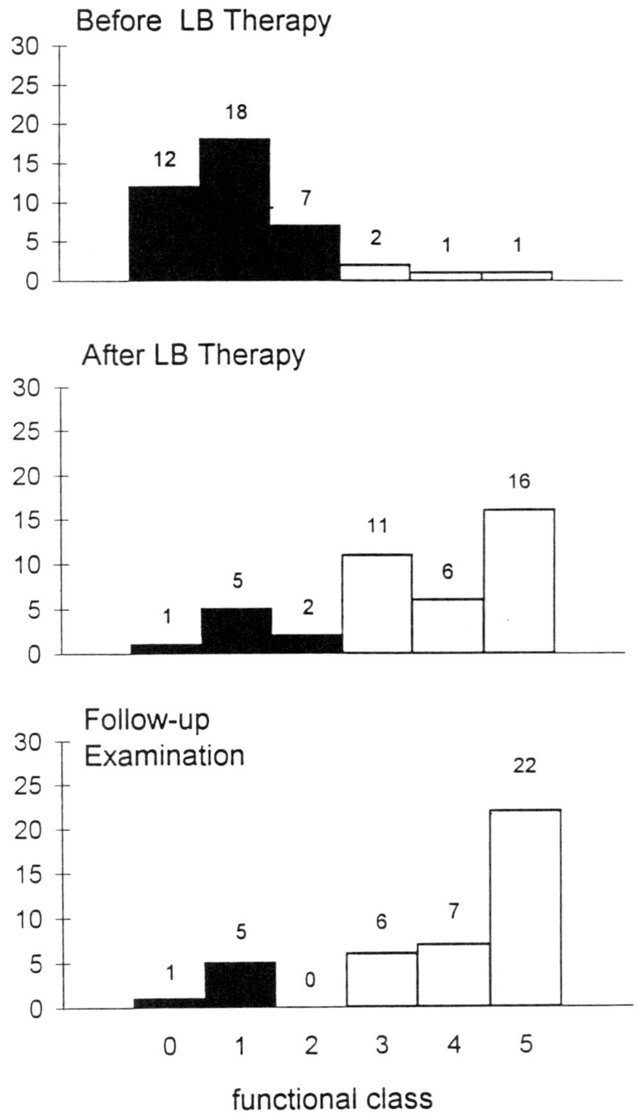

Figure 10.3. Walking capabilities according to 6 functional classes (0-5) in 41 acute para- and tetraplegic patients. Upper histogram: at the beginning of LB therapy during postacute rehabilitation 3 to 16 weeks (median 8 weeks) after spinal cord damage; middle histogram: the same patients at discharge from the hospital; lower histogram: follow-up investigations 0.5 to 6 years (median 17 months) after discharge from the hospital. Numbers on the columns indicated number of patients in this class.

Cumulative voluntary muscle activity (left and right limb)

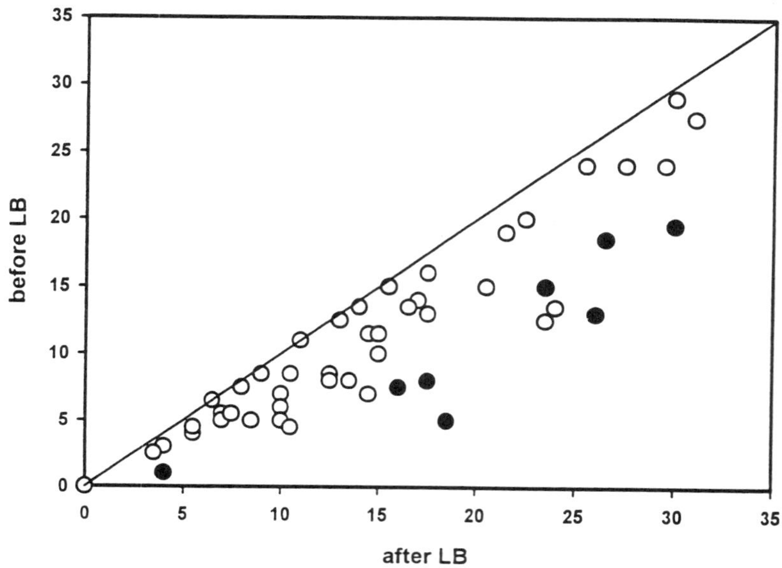

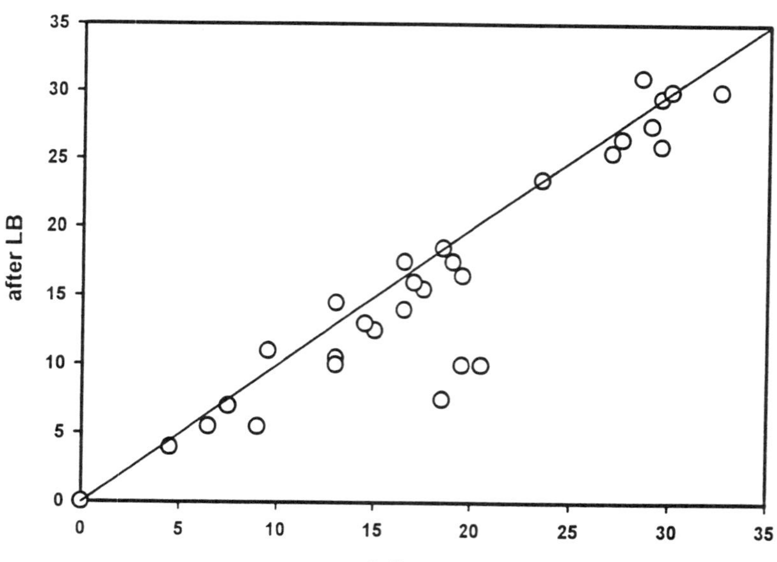

Figure 10.4. Upper diagram:Cumulative voluntary muscle activity of important flexor and extensor muscles of the limb bevor (ordinate) and after (abszissa) LB therapy of 25 initially wheelchair-bound chronic para- and tetraplegic patients. Black symbols: 5 patients with reduced antispastic medication.

In the light of the findings presented here, strategies of rehabilitation for incompletely paralyzed para- and tetraplegic persons need to focus on intensive training of upright walking. On the treadmill, body weight support by a harness is helpful, and the "rules of spinal locomotion" must be applied. Antispastic drugs reducing voluntary muscle activity need to be minimized. The safest strategy for the time being is to perform an intensive walking program with each patient in order to develop his/her maximum walking capability. Usually this is achieved within several weeks of daily training. The positive results of the follow-up investigations show that after achieving upright walking, the best exercise (even in patients with little voluntary muscle activity) is walking itself. The significant increase in voluntary muscle activity in 2 patients, years after the spinal cord injury (Figure 3 lower part), indicates that exercise dependent improvements may continue to occur. In many patients especially those with arm, shoulder, and rump paralyses the daily walking distance in the domestic surrounding is below the distance attained in a single 30 min training period on the Laufband (about 0.3 to 1.5 km). A continuing training on the Laufband after discharge from the hospital is useful therefore, at least for those severely paralyzed patients.

Since the successful application of LB-therapy in spinal cord injured persons, it has also been employed for patients after stroke (Richards, *et al.*,1993; Hesse, *et al.*,1994; Visintin, et al. 1998), and traumatic and hypoxic brain damage (A.Wernig, unpublished). It is quite obvious, that a harness supporting the patient to maintain upright position, the moving surface of the treadmill and the help with proper limb setting by therapists is a method suitable for training of patients with motor deficits of different causes.

10.6 REFERENCES

Barbeau, H., Blunt, R. (1991). A novel interactive locomotor approach using body weight support to retrain gate in spastic paretic subjects. In Wernig, A. (Ed.), *Plasticity of Motoneuronal Connections. Restorative Neurology Vol. 5* (pp. 461-474). Amsterdam: Elsevier.

Barbeau, H., Rossignol, S. Recovery of locomotion after chronic spinalization in the adult cat.*Brain Research,* 1987; 412:844-895.

Dietz, V., Colombo, G., Jensen, L., Baumgartner, L. Locomotor capacity of spinal cord in paraplegic patients. *Annals of Neurology,* 1995; 37: 574-586.

Dobkin, B.H., Harkema, S.J., Requejo, P.S., Edgerton, V.R. Modulation of locomotor-like EMG activity in subjects with complete and incomplete spinal cord injury. *Journal of Neurological Rehabilitation,* 1995; 9:183-190.

Edgerton, V.R., Roy, R.R., Hodgson, J.A., Gregor, R.J., de Guzman, C.P. (1991). Recovery of full weight-supporting locomotion of the hindlimbs after complete thoracic spinalization of adult and neonatal cats. In Wernig A. (Ed.), *Plasticity of Motoneuronal Connections. Restorative Neurology, Vol. 5* (pp. 405-418). Amsterdam: Elsevier.

Eidelberg, E., Walden, J.G., Nguyen, L.H. Locomotor control in macaque monkeys. *Brain,* 1981; 104:647-663.

Grillner, S. (1981). Control of locomotion in bipeds, tetrapods, and fish. In Brookhart, J.E., Mountcastle, V.B., Brooks, V.B., Geiger, S.R. (Eds.), *Handbook of Physiology, Section 1, Vol. 2, Part 2* (pp. 1127-1136). Bethesda, M.D.: American Physiological Society.

Harkema, S.J., Requejo, P.S., Hurley, S.L., Patel, U.K., Dobkin, B.H., Edgerton, V.R. Human lumbosacral spinal cord interprets loading during stepping. *Journal of Neurophysiology,* 1997; 77:797-811.

Harms, J. Screw-threaded rod system in spinal fusion surgery. *Spine: State of the Art Reviews,* 1992; 6:541-575.

Hesse, S.T., Bertelt, C., Schaffrin, A., Malezik, M., Mauritz, K.H. Restoration of gait in nonambulatory hemiparetic patients by treadmill training with partial body weight support. *Archives of Physical Medicine and Rehabilitation,* 1994; 75:1087-1093.

Kendall, H.O., Kendall, F.P., Wadsworth, G.E. *Muscles - Testing and Function.* Amsterdam: The Williams and Wilkins Company, 1971.

Lovely, R.G., Gregor, R.J., Roy, R.R., Edgerton, V.R. Effects of training on the recovery of full weight-bearing stepping in the spinal adult cat. *Experimental Neurology*, 1986; 92:421-435.

Maegele, M., Harkema, S.J., Requejo, P.S., Douglas, S.M., Müller, S., Wernig, A., Dobkin, B.H., Edgerton, V.R. EMG activity in lower limb muscles during voluntarilyand involuntarily generated movements with and without weight-bearing. *European Journal of Neuroscience Supplement*, 1996; 6:131.

Visintin, M., Barbeau, H. The effects of body weight support on the locomotor pattern of spastic paretic patients. *Canadian Journal of Neurological Sciences,* 1989; 16:315-325.

Wernig, A., Müller, S., Nanassy, A., Cagol, E. Laufband therapy based on "rules of spinal locomotion" is effective in spinal cord injured persons. *European Journal of Neuroscience,* 1995; 7:823-829.

Wernig, A., Nanassy, A., Müller, S. Maintainance of locomotor abilities following Laufband (treadmill) therapy in para- and tetraplegic persons: follow-up studies. *Spinal Cord,* 1998; 36:744-749.

Wernig, A., Müller, S. Improvement of walking in spinal cord injured persons after treadmill training. In: Plasticity of Motoneuronal Connections. *Restorative Neurology* Vol. 5, Wernig, A. (Ed), Amsterdam: Elsevier, 475-485, 1991.

Wernig, A., Müller, S. Laufband locomotion with body weight support improved walking in persons with spinal cord injuries. *Paraplegia,* 1992; 30:229-238.

Wernig, A., Müller, S. Die Lokomotionstherapie am Laufband bei Querschnittlähmung. Ergebnisse einer fünfjährigen Studie. *Journal of Neurological Rehabilitation,* 1995; 1:6-16.

Wernig, A., Müller, S. (1996) "Laufband" therapy based on the "rules of spinal locomotion" is effective in spinal cord injured persons. *European Journal of Neuroscience Supplement,* 1996; 9:57.

Subject Index

G

H

I

K

L

M

N

O

P